FINITE ELEMENT METHODS IN
STRUCTURAL MECHANICS

ELLIS HORWOOD SERIES IN MECHANICAL ENGINEERING

*In preparation

FINITE ELEMENT METHODS IN STRUCTURAL MECHANICS

C. T. F. ROSS, B.Sc. (Hons), Ph.D.
Reader in Applied Mechanics
Portsmouth Polytechnic

ELLIS HORWOOD LIMITED
Publishers · Chichester

Halsted Press: a division of
JOHN WILEY & SONS
New York · Chichester · Brisbane · Toronto

First published in 1985 by
ELLIS HORWOOD LIMITED
Market Cross House, Cooper Street, Chichester, West Sussex, PO19 1EB, England

The publisher's colophon is reproduced from James Gillison's drawing of the ancient Market Cross, Chichester.

Distributors:

Australia, New Zealand, South-east Asia:
Jacaranda-Wiley Ltd., Jacaranda Press,
JOHN WILEY & SONS INC.,
G.P.O. Box 859, Brisbane, Queensland 4001, Australia

Canada:
JOHN WILEY & SONS CANADA LIMITED
22 Worcester Road, Rexdale, Ontario, Canada.

Europe, Africa:
JOHN WILEY & SONS LIMITED
Baffins Lane, Chichester, West Sussex, England.

North and South America and the rest of the world:
Halsted Press: a division of
JOHN WILEY & SONS
605 Third Avenue, New York, N.Y. 10158 U.S.A.

©1985 C.T.F. Ross/Ellis Horwood Limited

British Library Cataloguing in Publication Data
Ross, C.T.F.
Finite element methods in structural mechanics. –
(Ellis Horwood series in mechanical engineering)
1. structures, Theory of 2. Finite element method
I. Title
624.1'71'01515353 TA646

Library of Congress Card No. 85–6180

ISBN 0–85312–808–1 (Ellis Horwood Limited – Library Edn.)
ISBN 0–85312–889–8 (Ellis Horwood Limited – Student Edn.)
ISBN 0–470–20207–6 (Halsted Press – Library Edn.)
ISBN 0–470–20208–4 (Halsted Press – Student Edn.)

Printed in Great Britain by Unwin Brothers of Woking

Table of Contents

Author's preface

Until very recently, the teaching of finite elements was aimed mostly at graduates, but with the advances made in computer technology, exposure to finite elements is now also at undergraduate level in many engineering institutions.

The book, therefore, is aimed at senior undergraduates, graduates, and engineers. Its purpose is to fill the gap between the numerous textbooks on traditional 'Strength of materials' and the postgraduate books that have been recently written on 'Finite element methods'. The topics are covered in the order the author has taught the subject to his students during the past twelve years.

Although most younger readers may already have received some tuition on matrix algebra, it is essential for the reader to be thoroughly familiar with this topic, prior to making any attempts to apply it to structural mechanics, and because of this, Chapter 1 has been included. The importance of this chapter is that it has been written by an engineer who specialises in this topic, rather than by a general mathematician, who may treat it as just another branch of mathematics. For this reason, the chapter lacks mathematical rigour, but encourages the reader to use matrix algebra as an engineering tool.

Chapter 2 covers the basic energy and stiffness concepts and introduces the matrix displacement method. Chapter 3 derives an elemental stiffness matrix for a rod and applies it to plane and space pin-jointed trusses.

Chapter 4 derives the slope-deflection equations for a beam and shows how these equations can be used to obtain the stiffness matrices for beams and columns. These elements are applied to beams and rigid-jointed plane and space frames with complex loading systems.

Chapter 5 introduces the finite element method proper, and derives elemental stiffness matrices for one- and two-dimensional elements. The two-dimensional elements include in-plane and out-of-plane plate elements and a 'shell' element. Thermal stress problems are also considered.

Chapter 6 extends the theory of Chapter 5 to the development of in-plane isoparametric quadrilateral elements. These elements are used to analyse plane stress and plane strain problems, via a microcomputer.

Chapter 7 further extends the theory of Chapter 5 to include vibration problems. Elemental mass matrices are derived for one and two-dimensional elements and 'shell' elements. Worked examples are given for the free vibration of a number of problems, including beams, plane frames, plates, and shells. Area coordinates are also introduced in this chapter.

Chapter 8 develops stiffness and mass matrices for a grillage element, and application is made to a number of grillages, via a computer.

Chapter 9 extends the theories of Chapters 5 and 7 to develop geometrical stiffness matrices for one- and two-dimensional elements. Application is then made to problems involving geometrical and material non-linearity and also nonlinear structural dynamics.

Most chapters include worked examples by hand and computer, and under the titles 'Examples for practice', at the end of most chapters, exercises are given for the reader to practise his/her newly acquired skills.

Computer programs
A number of the computer programs described in this book are available on tapes/disks for many microcomputers.

These computer programs can be purchased from:

> Ellis Horwood Limited, Publishers
> Department M
> Market Cross House
> Cooper Street
> Chichester
> West Sussex   PO19 1EB
> ENGLAND

Acknowledgements

The author would like to thank AUEW (TASS) of Richmond, Surrey, for permission to extract sections from the seven booklets he wrote for them.

His thanks are extended to his colleagues, Bill Barlow and Mac Lowcock; the former for his helpful comments and the latter for reading and checking the manuscript.

Finally, he would like to thank Mrs Lesley Jenkinson for the considerable care and patience she showed in typing the manuscript.

Introduction

Matrix methods of structural analysis first proved of interest in the late forties and the early fifties [1 to 3], as a result of a requirement for lighter aircraft structures.

The main problem at that time, however, was the lack or complete absence of computational power, as the high-speed digital computer with its own memory, was not invented until 1947. In fact, the first machine to be made commercially available, did not appear until 1951. This computer was called ENIAC, and according to Willis [4], it weighed about 30 tonnes, required about $110\,m^3$ to house it, and it consumed about $140\,kW$ of power. Furthermore, to prevent it from overheating, it required several tonnes of ice per day to keep it cool, and based on today's values, its purchase price was in millions of dollars!

Thus, it is evident that during that period, progress in matrix methods in structural mechanics was relatively slow, largely because of the lack of computational power. Indeed, during that period, teams of operators of desktop electromechanical calculators, often took several days to invert a matrix of modest size.

In 1956, a major breakthrough was made with numerical methods in structural mechanics, when Turner *et al* [5] invented the finite element method. Turner *et al.* showed how complex in-plane plate problems can be represented by finite elements of triangular shape, where each triangular element was described by three corner nodes.

Throughout the sixties, much progress was made with both computer technology and the development of more sophisticated finite elements [6 to 11].

In 1965, Melosh [12] realised that the finite element method could be extended to field problems by variational methods. This paper was an important contribution, as it led to a much wider use of the finite element method, and Zienkiewicz [13] applied it to a large number of steady-state and transient field problems.

At about the same time, researchers realised that a similar concept had been presented by Courant [14] in 1943. The main difference of Courant's concept to that of the finite element formulation, was that Courant used finite differences.

In the early 1960s, the integrated circuit was invented, and in 1970, Intel invented the first microprocessor. Since then, the accessibility to computational power has increased at an astonishing rate, and this has accelerated the use and development of the finite element method.

Today, computational power is much larger, more reliable, and relatively cheap, and as most technological concerns have access to computers, the popularity of using numerical methods is an ever increasing phenomenon. Even microcomputers [15, 16] can be used for finite element analysis, and as microcomputers become more powerful, it is likely that their effect will cause a technological explosion in the use of sophisticated methods for engineering science.

It is the author's belief that the effect of microcomputers will revolutionise the teaching of engineering science [17] during the eighties. Closed loop solutions will be replaced by numerical ones, wherever necessary, although if the problem is trivial and a closed loop trivial solution exists, then the engineer might as well use the simpler solution.

Finite elements require computers, for without computers the finite element method would be like trying to move an automobile without a motor.

In general, the finite element method is particularly useful for solving a differential equation, together with its boundary conditions, over a domain of complex shape. The process, therefore, is to represent the domain by a large number of finite elements of simpler shape, as shown in Fig. I.1. These finite elements are described by nodal points, the larger the number of nodes per element, the more sophisticated the element.

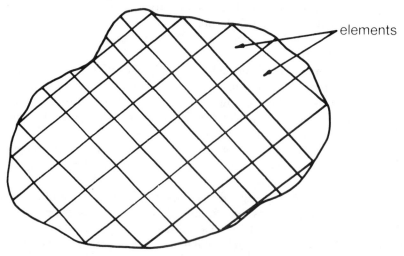

Fig. I.1 – Finite elements representing a domain.

By assuming an approximate variation of the required function over the finite element, and by considering elemental boundary conditions, the function approximation can be obtained in terms of the nodal values of the function for a particular element. Then, by considering equilibrium/compatibility at all the inter-element boundaries, together with known boundary conditions, a set of simultaneous equations will result.

Some of the simultaneous equations will be homogeneous and others non-homogeneous, but in general, their solution will give nodal values of the unknown function, together with other information.

Sometimes the nodal values of the function are all that are required, as in the case of temperature or Piezometric pressure head, but in other cases, such as in structural analysis, the function may be in the form of generalised displacements, hence it will usually be necessary to relate these generalised displacements to generalised forces, stresses, etc.

Generalised displacements take various forms, including translations and rotations, and these displacements correspond to generalised forces, such as line forces and couples, respectively.

Good FEM practice

In choosing an element, it is usually necessary to ensure that the theoretical predictions from the element, converge with an increase in mesh refinement.

One method of achieving this, is to select a simple problem, with known results, and to check the finite element by plotting its predictions against the number of nodes, as shown in Fig. I.2.

Fig. I.2 shows the results of a converging element, but non-converging elements, with results such as those of Fig. I.3, should be avoided, if possible.

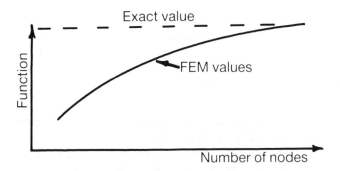

Fig. I.2 – Converging (conforming) element.

Other elements, which should be used with care, are those where the results only partially converge with increase in the number of nodes, as shown in Fig. I.4.

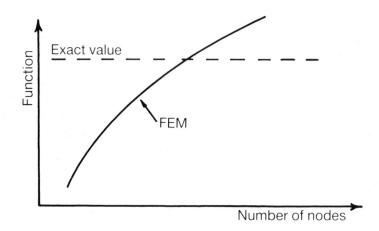

Fig. I.3 – Non-converging element.

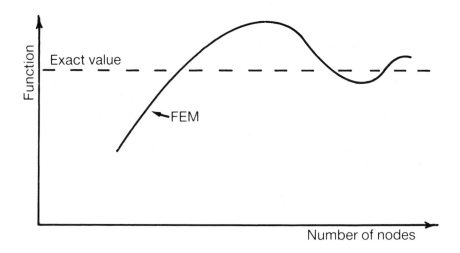

Fig. I.4 – Partially-converging element.

Another possible test on the converging properties of the element is the 'patch' test, as described by Irons [18, 19].

Even if the element is a converging one, care should be taken to avoid attaching a very stiff element to a very flimsy one, as shown in Fig. I.5. When structural members, such as in Fig. I.5, appear it is better to choose a mesh like that of Fig. I.6, where the variation in stiffness between any two adjacent elements is decreased. Such a choice usually improves numerical stability.

Numerical instability is a problem of which the engineer has to be continuously aware. Apart from numerical instability occurring due to a very stiff element being connected to a very flexible one, it can also occur if a badly

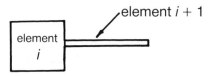

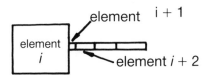

Fig. I.5 – Undesirable choice of elements.

Fig. I.6 – More desirable choice of elements.

shaped element is chosen, such as that shown in Fig. I.7. Ideally, triangular elements should be equilateral triangles and, in any case, the minimum angle between any two adjacent sides should not be less than 30°.

Fig. I.7 – Badly shaped triangular element.

Other badly shaped elements are those that can occur with the more sophisticated elements, as shown in Fig. I.8. In these cases, the elements have been badly distorted at the points shown by the arrows.

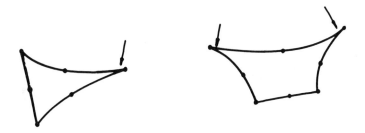

Fig. I.8 – Badly shaped elements.

Badly shaped elements very often cause negative numbers to appear on the main diagonal of the stiffness matrix, when, in fact, such numbers should always be positive.

Other numerical problems that occur with finite elements are the method of solution used and the number of equations to be solved. Either of these features can cause the numerical precision of the computer to be exceeded, and for many cases it will be necessary to resort to double precision arithmetic.

To guard against bad finite element practice, it will be necessary for the engineer to gain suitable experience from trial and error and also by searching the appropriate references [13, 19]. Perhaps the best test of all is to compare the finite element solution with large amounts of carefully obtained experimental observations.

Notation

Unless otherwise stated, the following symbols are adopted:

A	= cross-sectional area
I	= second moment of area
I_p	= polar moment of area
J	= torsional constant
l	= length
t	= thickness or time
T	= torque
M	= bending moment
n	= frequency (Hz)
r	= radius
R_1, R_2	= radii at nodes 1 and 2 respectively
E	= elastic modulus
G	= rigidity modulus
x, y, z	= coordinates (local axes)
$x^\mathrm{o}, y^\mathrm{o}, z^\mathrm{o}$	= coordinates (global axes)
X, Y, Z	= forces in x, y, and z directions respectively
$X^\mathrm{o}, Y^\mathrm{o}, Z^\mathrm{o}$	= forces in x^o, y^o, and z^o directions respectively
u, v, w	= displacements in x, y, and z directions respectively
$u^\mathrm{o}, v^\mathrm{o}, w^\mathrm{o}$	= displacements in x^o, y^o, and z^o directions respectively
α	= angle
λ	= eigenvalue
ω	= radian frequency
ρ	= density
σ	= stress
ε	= strain
τ_{xy}	= shear stress in the x–y plane
γ_{xy}	= shear strain in the x–y plane
v	= Poisson's ratio
γ	= $(1 - v)/2$ or $(1 - 2v)/[2(1 - v)]$

μ	$= v/(1-v)$
ξ	$= x/1$
$[k]$	= elemental stiffness matrix in local coordinates
$[k^0]$	= elemental stiffness matrix in global coordinates
$[k_G]$	= geometrical stiffness matrix in local coordinates
$[k_G{}^0]$	= geometrical stiffness matrix in global coordinates
$[m]$	= elemental mass matrix in local coordinates
$[m^0]$	= elemental mass matrix in global coordinates
$[K^0]$	= system stiffness matrix in global coordinates
$[K_G{}^0]$	= geometrical system stiffness matrix in global coordinates
$[M^0]$	= system mass matrix in global coordinates
$\{P_i\}$	= a vector of internal nodal forces
$\{q^0\}$	= a vector of external nodal forces in global coordinates
$\{u_i\}$	= a vector of nodal displacements in local coordinates
$\{u_i{}^0\}$	= a vector of nodal displacements in global coordinates
$[K_{11}]$	= that part of the system stiffness matrix that corresponds to the 'free' displacements
$[K_{G11}]$	= that part of the geometrical system stiffness matrix that corresponds to the 'free' displacements
$[M_{11}]$	= that part of the system mass matrix that corresponds to the 'free' displacements
$[C_v]$	= a matrix containing viscous damping terms
$[C_{11}]$	= that part of $[C_v]$ corresponding to the 'free' displacements
$[C_H]$	= a matrix containing hysterectic damping terms
$[\Xi]$	= a matrix of directional cosines
$[I]$	= identity matrix
$[\ \ ]$	= a square of rectangular matrix
$\{\ \ \}$	= a column vector
$\lfloor\ \ \rfloor$	= a row vector
$[0]$	= a null matrix

Parts of the Greek alphabet commonly used in mathematics

α	alpha
β	beta
γ	gamma
δ	delta
Δ	delta (capitals)
ε	epsilon
ζ	zeta
η	eta
θ	theta
κ	kappa
λ	lambda
μ	mu
v	nu

ξ	xi
Ξ	xi (capitals)
π	pi
σ	sigma
Σ	sigma (capitals)
τ	tau
ϕ	phi
χ	chi
ψ	psi
ω	omega
Ω	omega (capitals)

To Anne, Nicolette, Jonathan and mum, and to the memory of my late father

Chapter 1

Matrix Algebra

The approach in this chapter is based on technique rather than on rigorous mathematical theories. It commences with various matrix definitions, followed by the laws of matrix algebra. To demonstrate the latter, several examples are worked out in detail, and particular attention is paid to the inverse of a matrix and the solution of homogeneous and non-homogeneous simultaneous equations.

If the reader requires a greater depth of understanding of matrix algebra then he/she should study references [20 to 22].

1.1 DEFINITIONS

A *scalar* in its most usual form can be described as a number which is positive or negative or zero. Typical examples of scalars are 1, 2, π, e, -1.57, 2×10^{11}, etc., and typical scalar quantities appear in the form of temperature, time, mass, length, etc. Scalars have only magnitude.

A *vector* has both magnitude and direction, and typical vector quantities appear in the form of velocity, displacement, force, weight, etc.

A *matrix* in its most usual form is an array (or table) of scalar quantities, consisting of 'm' rows by 'n' columns, as shown in (1.1). The elements of the matrix need not necessarily be scalars, but can take the form of vectors or even matrices. This compact method of representing quantities, allows matrices to be particularly suitable for modelling physical problems on digital computers.

$$[A] = \begin{bmatrix} A_{11} & A_{12} & A_{13} & \cdot & \cdot & \cdot & \cdot & A_{1n} \\ A_{21} & A_{22} & A_{23} & \cdot & \cdot & \cdot & \cdot & A_{2n} \\ A_{31} & A_{32} & A_{33} & \cdot & \cdot & \cdot & \cdot & A_{3n} \\ \cdot & \cdot & \cdot & & & & & \cdot \\ \cdot & \cdot & \cdot & & & & & \cdot \\ \cdot & \cdot & \cdot & & & & & \cdot \\ A_{m1} & A_{m2} & A_{m3} & \cdot & \cdot & \cdot & \cdot & A_{mn} \end{bmatrix}. \tag{1.1}$$

A *row* of a matrix is defined as a horizontal line of quantities.

A *column* of a matrix is defined as a vertical line of quantities.

The quantities A_{11}, A_{12}, A_{13}, etc., are said to be the *elements* of the matrix $[A]$.

The *order* of a matrix is defined by its number of rows × its number of columns. Thus, the matrix of (1.1) is said to be of order $m \times n$.

A *column* matrix is where $n = 1$, as *in* (1.2).

$$\{A\} = \begin{Bmatrix} A_{11} \\ A_{21} \\ \cdot \\ \cdot \\ \cdot \\ A_{m1} \end{Bmatrix} \tag{1.2}$$

A *row matrix* is where $m = 1$, as in (1.3).

$$\lfloor A \rfloor = \lfloor A_{11} \quad A_{12} \quad \cdot \quad \cdot \quad \cdot \quad A_{1n} \rfloor \tag{1.3}$$

A *square matrix* is where $m = n$, as in (1.4).

$$[A] = \begin{bmatrix} A_{11} & A_{12} & \cdot & \cdot & \cdot & \cdot & A_{1n} \\ A_{21} & A_{22} & \cdot & \cdot & \cdot & \cdot & A_{2n} \\ \cdot & \cdot & & & & & \cdot \\ \cdot & \cdot & & & & & \cdot \\ \cdot & \cdot & & & & & \cdot \\ A_{n1} & A_{n2} & & & & & A_{nn} \end{bmatrix} \tag{1.4}$$

The square matrix of (1.4) is said to be of order n.

The *transpose* of a matrix is obtained by interchanging its rows with its columns, i.e. the transpose of a matrix is obtained by making its first column, its first row, and its second column, its second row, and so on and so forth. For example, if

$$[A] = \begin{bmatrix} 0 & -1 & 2 \\ 3 & 4 & -5 \end{bmatrix}$$

then the transpose of $[A]$ is given by

$$[A]^{\mathrm{T}} = \begin{bmatrix} 0 & 3 \\ -1 & 4 \\ 2 & -5 \end{bmatrix}.$$

A *super-matrix* is a matrix, whose elements themselves, are matrices, as shown in (1.5).

$$[A] = \begin{bmatrix} A_{11} & A_{12} & A_{13} & \cdot & \cdot & \cdot & \cdot & A_{1n} \\ A_{21} & A_{22} & A_{23} & & & & & A_{2n} \\ \hline A_{31} & A_{32} & A_{33} & & & & & A_{3n} \\ \cdot & \cdot & \cdot & & & & & \cdot \\ \cdot & \cdot & \cdot & & & & & \cdot \\ \cdot & \cdot & \cdot & & & & & \cdot \\ A_{m1} & A_{m2} & A_{m3} & \cdot & \cdot & \cdot & \cdot & A_{mn} \end{bmatrix}$$

$$= \left[\begin{array}{c|c} a & b \\ \hline c & d \end{array} \right] \tag{1.5}$$

where,

$$[a] = \begin{bmatrix} A_{11} & A_{12} \\ A_{21} & A_{22} \end{bmatrix}$$

$$[b] = \begin{bmatrix} A_{13} & \cdot & \cdot & \cdot & \cdot & \cdot & A_{1n} \\ A_{23} & \cdot & \cdot & \cdot & \cdot & \cdot & A_{2n} \end{bmatrix}$$

$$[c] = \begin{bmatrix} A_{31} & A_{32} \\ \cdot & \cdot \\ \cdot & \cdot \\ \cdot & \cdot \\ \cdot & \cdot \\ A_{m1} & A_{m2} \end{bmatrix}$$

$$[d] = \begin{bmatrix} A_{33} & \cdot & \cdot & \cdot & \cdot & \cdot & A_{3n} \\ \cdot & & & & & & \cdot \\ \cdot & & & & & & \cdot \\ \cdot & & & & & & \cdot \\ A_{m3} & & & & & & A_{mn} \end{bmatrix}.$$

The matrix of (1.5) is said to be partitioned, as shown by the broken lines. Matrix partitioning is found to be a very useful aid when isolating certain physical features within the matrix.

A *null* matrix is one which has all its elements equal to zero.

A *diagonal matrix* is a square matrix where all the elements except those of the main diagonal are zero, as in (1.6).

$$[A] = \begin{bmatrix} A_{12} & 0 & & & & & & 0 \\ 0 & A_{22} & & & & & & \\ 0 & 0 & A_{33} & & & & & \\ . & . & & & & & & \\ . & . & & & & & & \\ . & . & & & & & & \\ 0 & 0 & & & & & & A_{nn} \end{bmatrix}. \tag{1.6}$$

A *scalar matrix* is a diagonal matrix where all the diagonal elements are equal to the same scalar quantity. When the scalar quantity is unity, the matrix is called the *unit* or *identity matrix*, and is denoted by $[I]$.

An *upper triangular matrix* is a matrix which contains all its non-zero elements in and above its main diagonal, as in (1.7).

$$\begin{bmatrix} A_{11} & A_{12} & A_{13} & . & . & . & A_{1n} \\ 0 & A_{22} & A_{23} & . & . & . & A_{2n} \\ 0 & 0 & A_{33} & . & . & . & A_{3n} \\ . & . & . & & & & \\ . & . & . & & & & \\ . & . & . & & & & \\ 0 & 0 & 0 & & & & A_{nn} \end{bmatrix}. \tag{1.7}$$

A *lower triangular matrix* is one which contains all its non-zero elements in and below its main diagonal, as in (1.8).

$$\begin{bmatrix} A_{11} & 0 & . & . & . & . & 0 \\ A_{21} & A_{22} & & & & & \\ . & & . & & & & \\ . & & & . & & & \\ . & & & & . & & \\ . & & & & & & \\ A_{n1} & A_{n2} & . & . & . & . & A_{nn} \end{bmatrix} \tag{1.8}$$

A *band matrix* has all its non-zero elements contained in a diagonal strip as shown in (1.9) and (1.10). The centre diagonal of the strip is not necessarily the main diagonal.

$$\begin{bmatrix} A_{11} & A_{12} & 0 & 0 & 0 & 0 & . & . & . & . & 0 \\ A_{21} & A_{22} & A_{23} & 0 & 0 & 0 & . & . & . & . & 0 \\ 0 & A_{32} & A_{33} & A_{34} & 0 & & . & . & . & . & 0 \\ 0 & 0 & A_{43} & A_{44} & A_{45} & 0 & . & . & . & . & 0 \\ . & . & . & . & & & & & & & \\ . & . & . & . & . & & & & & & \\ . & . & . & . & . & . & & & & & \\ . & . & . & . & . & . & . & & & & \\ . & . & . & . & . & . & . & . & & & \\ . & . & . & . & . & . & . & . & . & & \\ 0 & 0 & 0 & . & 0 & & A_{n,\,n-1} & & A_{n,\,n} & \end{bmatrix} \qquad (1.9)$$

The bandwidth of the matrix of (1.10) is said to be NW.

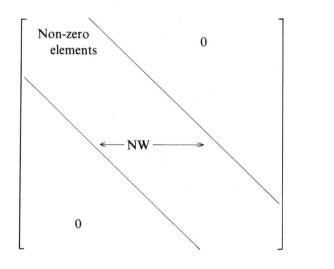

$$(1.10)$$

A *symmetric matrix* is where all

$$A_{ij} = A_{ji}$$

The *trace* of a matrix is obtained by summing all the elements on its leading diagonal, as follows:

$$\mathrm{Tr}\,[A] = \sum_{i=1}^{n} A_{ii},$$

and the *leading diagonal* of a square matrix consists of the elements: $A_{11}, A_{22}, A_{33}, \ldots, A_{nn}$.

1.2 ADDITION AND SUBTRACTION OF MATRICES

If

$$[A] = \begin{bmatrix} 1 & -1 \\ 2 & 3 \\ -4 & 5 \end{bmatrix}$$

and

$$[B] = \begin{bmatrix} -4 & 0 \\ -2 & 1 \\ 6 & -7 \end{bmatrix}$$

then,

$$[A] + [B] = \begin{bmatrix} (1-4) & (-1+0) \\ (2-2) & (3+1) \\ (-4+6) & (5-7) \end{bmatrix}$$

$$= \begin{bmatrix} -3 & -1 \\ 0 & 4 \\ 2 & -2 \end{bmatrix}$$

Similarly,

$$[A] - [B] = \begin{bmatrix} (1+4) & (-1-0) \\ (2+2) & (3-1) \\ (-4-6) & (5+7) \end{bmatrix}$$

$$= \begin{bmatrix} 5 & -1 \\ 4 & 2 \\ -10 & 12 \end{bmatrix}.$$

1.3 MATRIX MULTIPLICATION

In the relationship $[A][B] = [C]$, $[A]$ is known as the *premultiplier*, $[B]$ the *postmultiplier* and $[C]$ the *product*. Furthermore, if $[A]$ is of order $m \times n$ and $[B]$ is of order $n \times p$, then $[C]$ is of order $m \times p$. It should be noted that $[B]$ must always have its number of rows equal to the number of columns in $[A]$.

If

$$[A] = \begin{bmatrix} 1 & 2 \\ -1 & 0 \\ 3 & -4 \end{bmatrix}$$

and,

$$[B] = \begin{bmatrix} -2 & 3 & 0 \\ -1 & 4 & 1 \end{bmatrix},$$

then $[C]$ is obtained by multiplying the columns of $[B]$ by the rows of $[A]$, so that, in general,

$$C_{ij} = \sum_{k=1}^{n} A_{ik} \times B_{kj}$$

i.e, to obtain each C_{ij}, the ith row of $[A]$ must be premultiplied into the jth column of $[B]$, as follows:

$$[C] = \begin{bmatrix} 1 \times -2 + 2 \times (-1) & 1 \times 3 + 2 \times 4 & 1 \times 0 + 2 \times 1 \\ -1 \times -2 + 0 \times (-1) & -1 \times 3 + 0 \times 4 & -1 \times 0 + 0 \times 1 \\ 3 \times -2 - 4 \times (-1) & 3 \times 3 - 4 \times 4 & 3 \times 0 - 4 \times 1 \end{bmatrix}$$

$$= \begin{bmatrix} -4 & 11 & 2 \\ 2 & -3 & 0 \\ -2 & -7 & -4 \end{bmatrix}.$$

Similarly, if

$$\{A\} = \begin{Bmatrix} 1 \\ -2 \\ 3 \end{Bmatrix}$$

and

$$[B] = \lfloor -1 \quad 4 \quad 5 \rfloor$$

then

$$\{A\}\lfloor B \rfloor = \begin{bmatrix} 1 \times (-1) & 1 \times 4 & 1 \times 5 \\ -2 \times (-1) & -2 \times 4 & -2 \times 5 \\ 3 \times (-1) & 3 \times 4 & 3 \times 5 \end{bmatrix}$$

$$= \begin{bmatrix} -1 & 4 & 5 \\ 2 & -8 & -10 \\ -3 & 12 & 15 \end{bmatrix}$$

and

$$\lfloor B \rfloor \{A\} = (-1 \times 1) + (4 \times -2) + (5 \times 3) = -1 - 8 + 15 = 6.$$

Thus, in general, the vector product $\{A\}\lfloor B \rfloor$ will result in a matrix, and the product $\lfloor B \rfloor \{A\}$ will result in a scalar, and is sometimes called a *dot* product.

Furthermore, if

$$[A] = \begin{bmatrix} 4 & 1 & 2 \\ 5 & 0 & -6 \end{bmatrix}$$

and

$$[B] = \begin{vmatrix} 5 & 0 & 0 \\ 0 & 5 & 0 \\ 0 & 0 & 5 \end{vmatrix} \tag{1.11}$$

then

$$[A][B] = \begin{bmatrix} 20 & 5 & 10 \\ 25 & 0 & -30 \end{bmatrix}. \tag{1.12}$$

The result (1.12) can also be obtained from the expression $5 \times [A]$, and this is why the matrix of (1.11) is called a *scalar matrix*.

1.3.1   Some laws of matrix multiplication

$$([A][B])[C] = [A]([B][C])$$
$$[A]([B]+[C]) = [A][B]+[A][C]$$
$$[A][I] = [I][A]$$

If

$$[A][B] = [C][D]$$

then

$$[B]^{\mathrm{T}}[A]^{\mathrm{T}} = [D]^{\mathrm{T}}[C]^{\mathrm{T}}$$

If

$$[A] = [B][C]$$

then

$$[A]^{\mathrm{T}} = [C]^{\mathrm{T}}[B]^{\mathrm{T}}$$

Note, $[A][B] \neq [B][A]$ in general.

1.4   DETERMINANTS

The determinant of a square matrix $[A]$ is defined as,

$$|A| = \begin{vmatrix} A_{11} & A_{12} & . & . & . & . & A_{1n} \\ A_{21} & A_{22} & . & . & . & . & A_{2n} \\ . & . & . & & & & . \\ . & . & & & & & . \\ . & . & & & & & . \\ . & . & & & & & . \\ A_{n1} & A_{n2} & & & & & A_{nn} \end{vmatrix}.$$

The rule for expanding second order determinants is as follows:

$$|A| = \begin{vmatrix} A_{11} & A_{12} \\ A_{21} & A_{22} \end{vmatrix} \qquad\qquad (1.13)$$

so that

$$D = A_{11} A_{22} - A_{12} A_{21}$$

where

$$D = \text{the value of the determinant.}$$

Thus, the determinant 'D' of the matrix $|A|$ of (1.14) is evaluated as follows:

$$|A| = \begin{vmatrix} 4 & 2 \\ 3 & 6 \end{vmatrix} \qquad\qquad (1.14)$$

$$D = 4 \times 6 - 2 \times 3 = 18$$

The expression for expanding the second order determinant (1.13) can be extended to expand a third order determinant, as follows:
If

$$|A| = \begin{vmatrix} A_{11} & A_{12} & A_{13} \\ A_{21} & A_{22} & A_{23} \\ A_{31} & A_{32} & A_{33} \end{vmatrix}$$

then,

$$D = A_{11} \begin{vmatrix} A_{22} & A_{23} \\ A_{32} & A_{33} \end{vmatrix} - A_{12} \begin{vmatrix} A_{21} & A_{23} \\ A_{31} & A_{33} \end{vmatrix} + A_{13} \begin{vmatrix} A_{21} & A_{22} \\ A_{31} & A_{32} \end{vmatrix}$$

$$= A_{11} (A_{22} A_{33} - A_{23} A_{32}) - A_{12} (A_{21} A_{33} - A_{23} A_{31}) + A_{13} (A_{21} A_{32} - A_{22} A_{31}).$$

When D is zero, the matrix is said to be *singular*, and has no inverse (see Sec. 1.5).

Similar procedures can be adopted for higher order determinants, but these become extremely tedious with increase in size, and for these cases, the following methods are superior.

1.4.1 Chio's method

In this method, the elements in the leading diagonal are used as pivots, and from these positions, the elements below the leading diagonal are eliminated. By this process, the size of the determinant is decreased until eventually a 1×1 determinant is left.

The determinant (1.15) will be used to demonstrate the method.

Example 1.1

$$|A| = \begin{vmatrix} 2 & 0 & 1 \\ 4 & -2 & 3 \\ -1 & -3 & 5 \end{vmatrix}. \tag{1.15}$$

Divide the first row of (1.15) by 2 (the first pivot) and take this number outside the determinant, as shown in (1.16).

$$2 \begin{vmatrix} 1 & 0 & 0.5 \\ 4 & -2 & 3 \\ -1 & -3 & 5 \end{vmatrix}. \tag{1.16}$$

The value of the determinant of (1.16) is the same as that of (1.15).

To eliminate A_{21} from (1.16), multiply the first row of (1.16) by 4 and take it away from the second row, as follows:

$$2 \begin{vmatrix} 1 & 0 & 0.5 \\ (4-4) & (-2-0) & (3-2) \\ -1 & -3 & 5 \end{vmatrix}. \tag{1.17}$$

The value of the determinant of (1.17) is the same as (1.15) and (1.16).

To eliminate A_{31}, multiply the first row of (1.17) by (-1) and take it away from the third row of (1.17), as shown in (1.18).

$$= 2 \begin{vmatrix} 1 & 0 & 0.5 \\ 0 & -2 & 1 \\ 0 & -3 & 5.5 \end{vmatrix}. \tag{1.18}$$

Now divide A_{22} of (1.18) by -2 (the second pivot) and take this number outside the determinant, as shown in (1.19). The value of the determinant is not altered by this process.

$$2 \times (-2) \begin{vmatrix} 1 & 0 & 0.5 \\ 0 & 1 & -0.5 \\ 0 & -3 & 5.5 \end{vmatrix}. \tag{1.19}$$

To eliminate A_{32} from (1.19), multiply the second row of (1.19) by (-3) and take it away from the third row of (1.19).

$$2 \times (-2) \quad \begin{vmatrix} 1 & 0 & 0.5 \\ 0 & 1 & -0.5 \\ (0-0) & (-3+3) & (5.5-1.5) \end{vmatrix} \quad (1.20)$$

Equation (1.20) has now been reduced to an upper triangular matrix, so that its determinant is given by:

$$D = 2 \times (-2) \quad 1 \begin{vmatrix} 1 & -0.5 \\ 0 & 4 \end{vmatrix} - 0 \begin{vmatrix} 0 & -0.5 \\ 0 & 4 \end{vmatrix} + 0.5 \begin{vmatrix} 0 & 1 \\ 0 & 0 \end{vmatrix}$$

$$= -4(4-0+0) = \underline{-16}. \quad (1.21)$$

From (1.21) it can be seen that the determinant of a *triangular matrix* is given by the *products of its leading diagonal*.

Although the method of evaluating determinants by expansion is better than Chio's method for small determinants, the latter is superior for large determinants, as it lends itself to programming on a digital computer, as shown in Table 1.1

It should be noted that the method breaks down if a zero is met on the leading diagonal, but this can be overcome by interchanging two rows or two columns.

1.4.2 Choleski's method

Any positive-definite square matrix (see Sec. 1.7) can be represented as the product of an upper triangular matrix with a lower triangular matrix (known as *decomposition*), and as the determinant of a triangular matrix is simply the product of all the terms on its leading diagonal, the process can be very useful.

The determinant of (1.22) will be used to demonstrate the method.

Example 1.2

$$\begin{vmatrix} 2 & 6 \\ -2 & 4 \end{vmatrix}. \quad (1.22)$$

As (1.22) is not symmetrical, it will be necessary to make the elements of the leading diagonal of one of the triangular matrices equal to unity as follows:

$$\begin{vmatrix} 2 & 6 \\ -2 & 4 \end{vmatrix} = \begin{vmatrix} l_{11} & 0 \\ l_{21} & l_{22} \end{vmatrix} \times \begin{vmatrix} 1 & u_{12} \\ 0 & 1 \end{vmatrix}.$$

Table 1.1. Chio's method for evaluating a determinant.

```
100 REM CHIO'S METHOD OF EXPANDING DETERMINANTS
110 INPUT"TYPE IN ORDER OF DETERMINANT";N
120 DIM A(N,N)
130 PRINT"TYPE IN THE MATRIX"
140 FOR I=1 TO N
150 FOR J=1 TO N
160 PRINT"TYPE IN A(";I;",";J;")"
170 INPUT A(I,J)
180 NEXT J,I
190 DT=1
200 FOR I=1TO N-1
210 PI=A(I,I)
220 FOR J=I TO N
230 A(I,J)=A(I,J)/PI
240 NEXT J
250 FOR K=I TO N-1
260 CN=A(K+1,I)
270 FOR J=I TO N
280 A(K+1,J)=A(K+1,J)-A(I,J)*CN
290 NEXT J
300 NEXTK
310 DT=DT*PI
320 NEXT I
330 DT=DT*A(N,N)
340 PRINT"DETERMINANT=";DT
350 END
```

Equating coefficients, the following is obtained:

$$2 = l_{11} \times 1 + 0 \times 0$$

$$\underline{l_{11} = 2}$$

$$-2 = l_{21} \times 1 + l_{22} \times 0$$

$$\underline{l_{21} = -2}$$

$$6 = l_{11}u_{12} + 0 \times 1$$

$$u_{12} = \frac{6}{l_{11}} = \underline{3}$$

$$4 = l_{21}u_{12} + l_{22} \times 1$$

$$l_{22} = 4 - l_{21}u_{12} = 4 - (-2) \times 3$$

$$\underline{l_{22} = 10}$$

i.e.,

$$|L| = \begin{vmatrix} 2 & 0 \\ -2 & 10 \end{vmatrix}$$

and

$$|U| = \begin{vmatrix} 1 & 3 \\ 0 & 1 \end{vmatrix}.$$

$$D = \text{Det}|L| \times \text{Det}|U|$$
$$= (2 \times 10) \times (1 \times 1)$$
$$\underline{D = 20}.$$

This method is even more powerful for symmetrical matrices, and lends itself to computer analysis. Consider the symmetrical matrix of (1.23).

Example 1.3

$$\begin{bmatrix} 2 & -1 & 0 \\ -1 & 2 & -1 \\ 0 & -1 & 2 \end{bmatrix} \qquad (1.23)$$

As the matrix is symmetrical it can be represented by two triangular matrices, where $l_{ij} = l_{ji}$, as shown in (1.24).

$$\begin{bmatrix} 2 & -1 & 0 \\ -1 & 2 & -1 \\ 0 & -1 & 2 \end{bmatrix} = \begin{vmatrix} l_{11} & 0 & 0 \\ l_{21} & l_{22} & 0 \\ l_{31} & l_{32} & l_{33} \end{vmatrix} \times \begin{vmatrix} l_{11} & l_{12} & l_{13} \\ 0 & l_{22} & l_{23} \\ 0 & 0 & l_{33} \end{vmatrix} \qquad (1.24)$$

$$2 = l_{11}^2 + 0 + 0$$
$$\underline{l_{11} = 1.414}$$

$$-1 = l_{11}l_{12} + 0 \times l_{22} + 0$$
$$\underline{l_{12} = -0.707}$$

$$0 = l_{11} \times l_{13} + 0 \times l_{23} + 0 \times l_{33}$$
$$\underline{l_{13} = 0}$$

$$2 = l_{21}l_{12} + l_{22}^2 + 0 \times 0$$
$$l_{22}^2 = -l_{12}^2 + 2, \text{ therefore } \underline{l_{22} = 1.2247}$$

$$-1 = l_{21}l_{13} + l_{22}l_{23} + 0 \times l_{33}$$
$$-1 = 0 + 1.2247\, l_{23} + 0$$
$$\underline{l_{23} = -0.8165}$$

$$2 = l_{31} \, l_{13} + l_{32} \, l_{23} + l_{33}^2$$
$$l_{33}^2 = 2 - 0 - (0.667)$$
$$l_{33} = 1.155$$

$$[L] = \begin{bmatrix} 1.414 & 0 & 0 \\ -0.707 & 1.2247 & 0 \\ 0 & -0.8165 & 1.155 \end{bmatrix}$$

$$\text{Det} = (1.414 \times 1.2247 \times 1.155)^2$$
$$= 4.$$

This method, too, can be seen to be favourable for solution by computer.

1.4.3 Minors and cofactors

The *minor* of a determinant is a smaller determinant within the original determinant, and is formed by removing from the latter, an equal number of rows and columns. For example, if the determinant is

$$\begin{vmatrix} A_{11} & A_{12} & A_{13} \\ A_{21} & A_{22} & A_{23} \\ A_{31} & A_{32} & A_{33} \end{vmatrix} \tag{1.25}$$

then some of its second order minors can be obtained by deleting the first row, and the third, first, and second columns respectively, to give the following

$$\begin{vmatrix} A_{21} & A_{22} \\ A_{31} & A_{32} \end{vmatrix} \quad \begin{vmatrix} A_{22} & A_{23} \\ A_{32} & A_{33} \end{vmatrix} \quad \begin{vmatrix} A_{21} & A_{23} \\ A_{31} & A_{33} \end{vmatrix}$$

Other minors can be obtained by removing the second row and each of the three columns, and then the process can be repeated for the third row to give yet another set of minors.

The *cofactor* of some elements of the determinant of (1.25) can be found as follows:

$$A_{11}^{c} = \begin{vmatrix} A_{22} & A_{23} \\ A_{32} & A_{33} \end{vmatrix}$$

$$A_{12}^{c} = - \begin{vmatrix} A_{21} & A_{23} \\ A_{31} & A_{33} \end{vmatrix} \tag{1.26}$$

$$A_{13}^{c} = \begin{vmatrix} A_{21} & A_{22} \\ A_{31} & A_{32} \end{vmatrix}$$

$$A_{21}^{c} = - \begin{vmatrix} A_{12} & A_{13} \\ A_{32} & A_{33} \end{vmatrix}.$$

In general, for an nth order determinant,

$$A_{ij}{}^c = (-1)^{i+j} \begin{vmatrix} A_{11} & A_{12} & A_{1\,j-1} & A_{1,\,j+1} & \cdot & \cdot & \cdot & A_{1n} \\ A_{21} & A_{22} & A_{2,\,j-1} & A_{2,\,j+1} & \cdot & \cdot & \cdot & A_{2n} \\ & & \cdot & \cdot & \cdot & \cdot & \cdot & \cdot \\ & \cdot & \cdot & & \cdot & \cdot & \cdot & \cdot \\ A_{i-1,1} & A_{i-1,2} & \cdot & \cdot & \cdot & \cdot & \cdot & \cdot \\ A_{i+1,1} & A_{i+1,2} & \cdot & \cdot & \cdot & \cdot & \cdot & \cdot \\ & \cdot & \cdot & \cdot & \cdot & \cdot & & \cdot \\ A_{n1} & & & & & & & A_{nn} \end{vmatrix} \qquad (1.27)$$

1.4.4 Cofactor matrix

This is obtained by replacing each element of a matrix by its cofactor, so that

$$[A]^c = \begin{bmatrix} A_{11}{}^c & A_{12}{}^c & A_{13}{}^c & : & : & : & A_{1n}{}^c \\ A_{21}{}^c & A_{22}{}^c & & & \cdot & \cdot & A_{2n}{}^c \\ \cdot & & & & & & \\ \cdot & \cdot & & & & & \\ \cdot & \cdot & & & & & \\ A_{n1}{}^c & A_{n2}{}^c & & & \cdot & \cdot & A_{nn}{}^c \end{bmatrix}.$$

Adjoint matrix
This is merely the transpose of the cofactor matrix.

$$[A]^a = [A]^{cT}.$$

1.5 MATRIX INVERSE

Determining the inverse of a matrix is analogous to finding the reciprocal of a scalar. For this reason, an inverted matrix is sometimes called a *reciprocal matrix*, and the inverse of a matrix $[A]$ is denoted by $[A^{-1}]$. Only square non-singular matrices can be inverted.

In its simplest form the inverse of a matrix is obtained from expression (1.28):

$$[A]^{-1} = \frac{[A]^a}{|A|}. \qquad (1.28)$$

Example 1.4
Using expression (1.28), determine the inverse of (1.29).

$$[A] = \begin{bmatrix} 2 & 4 & 3 \\ 1 & -2 & 0 \\ -1 & -4 & 5 \end{bmatrix}. \qquad (1.29)$$

The first step will be to obtain the cofactors, so equations (1.27) will be adopted:

$$A_{11}{}^c = \begin{vmatrix} -2 & 0 \\ -4 & 5 \end{vmatrix} = -10$$

$$A_{12}{}^c = -\begin{vmatrix} 1 & 0 \\ -1 & 5 \end{vmatrix} = -5$$

$$A_{13}{}^c = \begin{vmatrix} 1 & -2 \\ -1 & -4 \end{vmatrix} = -6$$

$$A_{21}{}^c = -\begin{vmatrix} 4 & 3 \\ -4 & 5 \end{vmatrix} = -32.$$

Similarly for the other cofactors of (1.29), so that,

$$|A^c| = \begin{vmatrix} -10 & -5 & -6 \\ -32 & 13 & 4 \\ 6 & 3 & -8 \end{vmatrix}.$$

Now,

$$|A| = 2 \times A_{11}{}^c + 4 \times A_{12}{}^c + 3 \times A_{13}{}^c$$
$$= -20 - 20 - 18 = -58$$

therefore

$$[A^{-1}] = \begin{bmatrix} 0.1724 & 0.5517 & -0.1034 \\ 0.0862 & -0.2241 & -0.0517 \\ 0.1034 & -0.0689 & 0.1379 \end{bmatrix}.$$

Example 1.5
Determine a general expression for the inverse of the 2×2 matrix of (1.30).

$$[A] = \begin{bmatrix} A_{11} & A_{12} \\ A_{21} & A_{22} \end{bmatrix} \qquad\qquad (1.30)$$

$$A_{11}{}^c = A_{22}$$
$$A_{12}{}^c = -A_{21}$$
$$A_{21}{}^c = -A_{12}$$
$$A_{22}{}^c = A_{11}.$$

Hence,

$$[A]^c = \begin{bmatrix} A_{22} & -A_{21} \\ -A_{12} & A_{11} \end{bmatrix}.$$

Now,

$$|A| = A_{11} \times A_{22} - A_{12} A_{21}$$

therefore

$$[A^{-1}] = \frac{\begin{bmatrix} A_{22} & -A_{12} \\ -A_{21} & A_{11} \end{bmatrix}}{A_{11} \times A_{22} - A_{12} A_{21}}.$$ (1.31)

Expression (1.28) is a satisfactory method of inverting small matrices, but becomes a cumbersome expression for inverting larger matrices, the time taken to invert a matrix increasing with the cube of its order. Thus, for larger matrices, it is desirable to invert matrices by computer, and some of these methods are described below.

1.5.1 Method of successive transformations (Gauss-Jordan)

This is a pivotal method and is somewhat similar to Chio's method for evaluating determinants.

Simultaneous equations are sometimes solved by using the property of an inverse matrix, and this principle is used to formulate the process for the present method.

Consider the following set of simultaneous equations:

$$A_{11} x_1 + A_{12} x_2 + .. + A_{1n} x_n = C_1$$
$$A_{21} x_1 + A_{22} x_2 + .. + A_{2n} x_n = C_2$$

$$. \qquad . \qquad . $$
$$. \qquad . \qquad . $$
$$. \qquad . \qquad . $$

$$A_{n1} x_1 + A_{n2} x_2 + .. + A_{nn} x_n = C_n.$$ (1.32)

In matrix form equations (1.32) can be rewritten as follows:

$$[A] \{x\} = [C].$$ (1.33)

Premultiplying both sides of (1.33) by $[A]^{-1}$, the following is obtained:

$$[A]^{-1} [A] \{x\} = [A]^{-1} \{C\}.$$ (1.34)

Now, as the definition of the inverse of a matrix is

$$[A]^{-1} [A] = [A] [A]^{-1} = \lceil I \rfloor,$$

equation (1.34) becomes

$$\{x\} = [A]^{-1} \{C\}.$$ (1.35)

Similarly (1.32) can be put in the form:

$$
\begin{bmatrix}
A_{11} & A_{12} & . & . & A_{1n} & 1 & 0 & . & . & . & 0 \\
A_{21} & A_{22} & . & . & A_{2n} & 0 & 1 & & & & 0 \\
. & . & . & & & . & & . & & \\
. & . & . & & & . & & & . & \\
. & . & . & & & . & & & & . \\
. & . & . & & & . & & & . & \\
A_{n1} & A_{n2} & . & . & A_{nn} & 0 & . & . & . & . & 1
\end{bmatrix}
\begin{Bmatrix}
x_1 \\ x_2 \\ . \\ . \\ . \\ x_n \\ -C_1 \\ C_2 \\ . \\ . \\ . \\ . \\ -C_n
\end{Bmatrix}
= \{0\}
\tag{1.36}
$$

If in equation (1.36), the matrix $[A]$ is transformed into a Unit matrix and the operations necessary to do this are also carried out on the Unit matrix of (1.36), the following is obtained:

$$
[I|B] \left\{ \frac{x}{-C} \right\} = \{0\},
\tag{1.37}
$$

therefore

$$\{x\} = [B]\{C\}$$

Comparison of (1.37) with (1.35) shows that

$$[B] = |A|^{-1}.$$

Example 1.6
Invert the matrix of (1.29) by the method of successive transformations. The matrix must first be written in the form:

$$
\begin{bmatrix}
2 & 4 & 3 & 1 & 0 & 0 \\
1 & -2 & 0 & 0 & 1 & 0 \\
-1 & -4 & 5 & 0 & 0 & 1
\end{bmatrix}.
\tag{1.38}
$$

Divide the first row of (1.38) by A_{11} (the first pivot) to give (1.39):

$$
\begin{bmatrix}
1 & 2 & 1.5 & 0.5 & 0 & 0 \\
1 & -2 & 0 & 0 & 1 & 0 \\
-1 & -4 & 5 & 0 & 0 & 1
\end{bmatrix}.
\tag{1.39}
$$

Multiply the 1st row of (1.39) by A_{21}, which is one in this case, and take it away from the 2nd row to give (1.40):

$$
\left[
\begin{array}{ccc|ccc}
1 & 2 & 1.5 & 0.5 & 0 & 0 \\
0 & -4 & -1.5 & -0.5 & 1 & 0 \\
-1 & -4 & 5 & 0 & 0 & 1
\end{array}
\right]. \tag{1.40}
$$

Multiply the 1st row of (1.40) by A_{31}, (-1), and take it away from the 3rd row of (1.40) to give (1.41):

$$
\left[
\begin{array}{ccc|ccc}
1 & 2 & 1.5 & 0.5 & 0 & 0 \\
0 & -4 & -1.5 & -0.5 & 1 & 0 \\
0 & -2 & 6.5 & 0.5 & 0 & 1
\end{array}
\right]. \tag{1.41}
$$

Divide the 2nd row of (1.41) by A_{22}, (-4), (the second pivot) to give (1.42):

$$
\left[
\begin{array}{ccc|ccc}
1 & 2 & 1.5 & 0.5 & 0 & 0 \\
0 & 1 & 0.375 & 0.125 & 0.25 & 0 \\
0 & -2 & 6.5 & 0.5 & 0 & 1
\end{array}
\right]. \tag{1.42}
$$

To eliminate A_{32} multiply the 2nd row of (1.42) by A_{32}, (-2), and take it away from the 3rd row.

$$
\left[
\begin{array}{ccc|ccc}
1 & 2 & 1.5 & 0.5 & 0 & 0 \\
0 & 1 & 0.375 & 0.125 & 0.25 & 0 \\
0 & 0 & 7.25 & 0.75 & 0.5 & 1
\end{array}
\right]. \tag{1.43}
$$

Divide the 3rd row of (1.43) by 7.25 (the third pivot) to give (1.44):

$$
\left[
\begin{array}{ccc|ccc}
1 & 2 & 1.5 & 0.5 & 0 & 0 \\
0 & 1 & 0.375 & 0.125 & 0.25 & 0 \\
0 & 0 & 1 & 0.1034 & 0.06897 & 0.1379
\end{array}
\right]. \tag{1.44}
$$

To eliminate A_{23}, multiply the 3rd row of (1.44) by A_{23} and take it away from the 2nd row.

$$
\left[
\begin{array}{ccc|ccc}
1 & 2 & 1.5 & 0.5 & 0 & 0 \\
0 & 1 & 0 & 0.0862 & 0.2241 & -0.0517 \\
0 & 0 & 1 & 0.1034 & 0.06897 & 0.1379
\end{array}
\right]. \tag{1.45}
$$

To eliminate A_{12} from (1.45), multiply the 3rd row of (1.45) by 1.5 and take it away from the 1st row of (1.45):

$$
\begin{bmatrix}
1 & 2 & 0 & 0.3449 & -0.1034 & -0.2069 \\
0 & 1 & 0 & 0.0862 & 0.2241 & -0.0517 \\
0 & 0 & 1 & 0.1034 & 0.06897 & 0.1379
\end{bmatrix}. \tag{1.46}
$$

To eliminate A_{12} from (1.46), multiply the 2nd row by 2, and take it away from the 1st row to give (1.47):

$$
\begin{bmatrix}
1 & 0 & 0 & 0.1725 & -0.5516 & -0.1035 \\
0 & 1 & 0 & 0.0862 & -0.2241 & -0.0517 \\
0 & 0 & 1 & 0.1034 & -0.0690 & 0.1379
\end{bmatrix}. \tag{1.47}
$$

Hence, from (1.47) it can be seen that

$$
[A^{-1}]
\begin{bmatrix}
0.1725 & -0.5516 & -0.1035 \\
0.0862 & -0.2241 & -0.0517 \\
0.1034 & -0.0690 & 0.1379
\end{bmatrix}. \tag{1.48}
$$

The matrix of (1.48) can be seen to compare favourably with the inverse obtained from expression (1.28), and readers who are familiar with computer programming can see that this method lends itself readily to analysis by computer.

The computer program of Table 1.2 is based on this method, but the unit matrix is not stored, as its properties are known.

It should be noted, however, that the method breaks down if a zero is encountered on the leading diagonal; but this difficulty can be overcome by interchanging *two rows* or two columns, and after the matrix has been inverted, the corresponding *two columns* or rows must be interchanged.

1.5.2 Choleski's method

Although this method can be used for inverting symmetrical and unsymmetrical matrices, only the former will be considered, as the process is more suitable for these cases.

As stated in section 1.4.2, any positive-definite square matrix can be represented by the product of an upper triangular matrix with a lower triangular matrix, and if the original matrix is *symmetrical*, the upper triangular matrix will be the transpose of the lower triangular matrix.

Let this lower triangular matrix be

$$
[L] =
\begin{bmatrix}
l_{11} & 0 & . & . & . & . & 0 \\
l_{21} & l_{22} & . & . & . & . & 0 \\
. & . & . & . & . & . & . \\
. & . & . & . & . & . & . \\
. & . & . & . & . & . & . \\
. & . & . & . & . & . & . \\
l_{n1} & l_{n2} & . & . & . & . & l_{nn}
\end{bmatrix} \tag{1.49}
$$

Table 1.2. Program for the inverse of an unsymmetrical matrix.

```
5 PRINT"⊐"
10 PRINT:PRINT:PRINT"INVERSE OF A REAL UNSYMMETRICAL MATRIX OF NTH ORDER (IN SI
TU)"
11 PRINT:PRINT:PRINT"THE METHOD IS BASED ON THAT OF GAUSS-JORDAN":PRINT:PRINT
12 PRINT"IT BREAKS DOWN IF THERE IS A ZERO ON THE MAIN DIAGONAL"
20 PRINT:PRINT:PRINT "FEED IN ORDER OF MATRIX"
30 INPUT N
35 DIM A(N,N)
40 PRINT "FEED IN THE MATRIX TO BE INVERTED"
50 FOR II=1 TO N
60 FOR JJ=1 TO N
70 PRINT "ROW NO";II
80 PRINT "COLUMN NO";JJ
90 INPUT A(II,JJ)
100 NEXT JJ
110 NEXT II
120 FOR M=1 TO N
130 DI=A(M,1)
140 IF DI=0 THEN PRINT "THE MATRIX IS SINGULAR"
150 FOR Q=1 TO N-1
160 A(M,Q)=A(M,Q+1)/DI
170 NEXT Q
180 A(M,N)=1/DI
190 FOR P=1 TO N
200 IF P=M THEN 260
210 R=A(P,1)
220 FOR Q=1 TO N-1
230 A(P,Q)=A(P,Q+1)-R*A(M,Q)
240 NEXT Q
250 A(P,N)=-R*A(M,N)
260 NEXT P
270 NEXT M
275 PRINT:PRINT:PRINT"THE INVERSE IS AS FOLLOWS":PRINT:PRINT
280 FOR II=1 TO N
290 FOR JJ=1 TO N
300 PRINT A(II,JJ);
310 NEXT JJ
320 PRINT
330 NEXT II
340 PRINT:PRINT
350 END
```

so that

$$[A] = [L][L]^T. \tag{1.50}$$

Equating coefficients in (1.50), the following are obtained for the elements of $[L]$.

$$l_{ij} = 0 \text{ for } i < j$$

$$l_{ii} = \left(A_{ii} - \sum_{r=1}^{i-1} l_{ir}^2 \right)^{\frac{1}{2}} \tag{1.51}$$

$$l_{ij} = \left(A_{ij} - \sum_{r=1}^{j-1} l_{ir} l_{jr} \right) \Big/ l_{jj} \text{ for } i > j.$$

Now it can be proved that the inverse of a lower triangular matrix is another lower triangular matrix and vice versa for an upper triangular matrix. Let

$$[B] = [L]^{-1}$$

so that

$$
\begin{bmatrix}
l_{11} & 0 & . & . & . & . & . & 0 \\
l_{21} & l_{22} & & & & & & \\
. & & . & & & & & \\
. & & & . & & & & \\
. & & & & . & & & \\
. & & & & & . & & \\
l_{n1} & l_{n2} & . & . & . & . & . & l_{nn}
\end{bmatrix}
\begin{bmatrix}
b_{11} & 0 & . & . & . & . & . & 0 \\
b_{21} & b_{22} & & & & & & \\
. & & . & & & & & \\
. & & & . & & & & \\
. & & & & . & & & \\
. & & & & & . & & \\
b_{n1} & b_{n2} & . & . & . & . & . & b_{nn}
\end{bmatrix}
= |I|.
$$

(1.52)

Equating coefficients in (1.52), the following expressions are obtained for the elements of $[B]$.

$$b_{ij} = 0 \text{ for } i < j$$
$$b_{ii} = 1/l_{ii}$$
$$b_{ij} = -\left(\sum_{r=j}^{i-1} l_{ir} b_{rj}\right)\bigg/ l_{ii} \text{ for } i > j. \tag{1.53}$$

Hence, the inverse of $[A]$ can be found from (1.54):

$$[A]^{-1} = ([L][L]^{\mathrm{T}})^{-1} = [B]^{\mathrm{T}}[B]. \tag{1.54}$$

Example 1.7
Invert the symmetrical matrix of (1.55) by the Choleski method.

$$
\begin{bmatrix}
2 & 1 & 1 \\
1 & 1.5 & 2 \\
1 & 2 & 6.75
\end{bmatrix}.
\tag{1.55}
$$

First, (1.55) must be represented by the product of an upper and lower triangular matrix (known as a *Choleski decomposition*).
From (1.51),

$$l_{12} = l_{13} = l_{23} = 0$$
$$l_{11} = \sqrt{2} = 1.414$$
$$l_{21} = (1-0)/1.414 = 0.707$$
$$l_{31} = (1-0)/1.414 = 0.707$$
$$l_{22} = (1.5 - 0.707^2)^{1/2} = 1$$
$$l_{32} = (2 - 0.707^2)/1 = 1.5$$
$$l_{33} = (6.75 - (0.707^2 + 1.5^2))^{1/2} = (6.75 - 2.75)^{1/2} = 2.0.$$

So that,

$$[L] = \begin{bmatrix} 1.414 & 0.0 & 0.0 \\ 0.707 & 1.0 & 0.0 \\ 0.707 & 1.5 & 2.0 \end{bmatrix}. \tag{1.56}$$

From (1.53) and (1.56)

$$b_{12} = b_{13} = b_{23} = 0$$
$$b_{11} = 1/1.414 = 0.707$$
$$b_{21} = -(0.707 \times 0.707)/1 = -0.5$$
$$b_{31} = -(0.707 \times 0.707 + 1.5 \times -0.5)/2$$
$$= -(0.5 - 0.75)/2 = 0.125$$
$$b_{22} = 1$$
$$b_{32} = -(1.5 \times 1)/2 = -0.75$$
$$b_{33} = 0.5.$$

So that

$$[B] = \begin{bmatrix} 0.707 & 0.0 & 0.0 \\ -0.5 & 1.0 & 0.0 \\ 0.125 & -0.75 & 0.5 \end{bmatrix}. \tag{1.57}$$

and

$$[B]^{\mathrm{T}} = \begin{bmatrix} 0.707 & -0.5 & 0.125 \\ 0.0 & 1.0 & -0.75 \\ 0.0 & 0.0 & 0.5 \end{bmatrix}. \tag{1.58}$$

Substituting (1.57) and (1.58) into (1.54), the inverse of $[A]$ is as shown in (1.59):

$$[A]^{-1} = \begin{bmatrix} 0.77 & -0.594 & 0.063 \\ -0.594 & 1.563 & -0.375 \\ 0.063 & -0.375 & 0.25 \end{bmatrix}. \tag{1.59}$$

It should be noted that the inverse of a symmetrical matrix is also symmetrical.

1.5.3 Method of improving the accuracy of an inverted matrix

This method is suitable for diagonally dominant matrices, such as those met in structural mechanics.

Let

$$[A]^{-1} = \text{a close approximation to the inverse of the matrix.}$$

and

$$[\Delta A_1] = \text{a matrix of corrections to } [A_1]^{-1}$$

so that,

$$[A]^{-1} = [A_1]^{-1} + [\Delta A_1] \tag{1.60}$$

Premultiplying both sides of (1.60) by $[A_1]^{-1} [A]$, the following is obtained:

$$[A_1]^{-1} = [A_1]^{-1} [A] [A_1]^{-1} + [A_1]^{-1} [A] [\Delta A_1] \tag{1.61}$$

Assuming that

$$[A_1]^{-1} [A] \simeq [I],$$

then from (1.61)

$$[\Delta A_1] \simeq [A_1]^{-1} ([I] - [A] [A_1]^{-1}),$$

but

$$[\Delta A_1] = [A]^{-1} - [A_1]^{-1}$$

therefore

$$[A]^{-1} - [A_1]^{-1} = [A_1]^{-1} ([I] - [A] [A_1]^{-1}),$$

i.e. $$[A]^{-1} = [A_1]^{-1} (2[I] - [A] [A_1]^{-1})$$

Hence, for the nth approximation,

$$[A_n]^{-1} = [A_{n-1}]^{-1} (2[I] - [A] [A_{n-1}])^{-1}. \tag{1.62}$$

1.5.4 Orthogonal matrices

In engineering, these are usually met when certain physical properties, which are orthogonal, are transformed from one set of cartesian coordinates to another. Typical examples of such cases are two-dimensional stress and strain systems and principal second moments of area.

In tensor form, the relationship between two sets of orthogonal stresses can be represented as shown in (1.63):

$$\begin{pmatrix} \sigma_\theta & \tau_\theta \\ \tau_\theta & \sigma_{90+\theta} \end{pmatrix} = \begin{bmatrix} c & s \\ -s & c \end{bmatrix} \begin{pmatrix} \sigma_x & \tau_{xy} \\ \tau_{xy} & \sigma_y \end{pmatrix} \begin{bmatrix} c & -s \\ s & c \end{bmatrix} \tag{1.63}$$

where

$$\sigma_\theta = \text{direct stress at any angle } \theta$$
$$\sigma_{90+\theta} = \text{direct stress at any angle } 90 + \theta$$

τ_θ = shear stress in $\theta - (90 + \theta)$ plane

σ_x = direct stress in x direction

σ_y = direct stress in y direction

τ_{xy} = shear stress in x–y plane

θ = angle with x axis

$c = \cos\theta$

$s = \sin\theta$

Expansion of (1.63) gives the following well-known expressions {23}:

$$\sigma_\theta = \sigma_x c^2 + \sigma_y s^2 + 2\tau_{xy}cs$$
$$\sigma_{90+\theta} = \sigma_x s^2 + \sigma_y c^2 - 2\tau_{xy}cs$$
$$\tau_\theta = -\sigma_x cs + \sigma_y cs + (c^2 - s^2)\tau_{xy}.$$

From (1.63), it can be seen that the matrices relating these two systems are as follows:

$$[\zeta] = \begin{bmatrix} c & s \\ -s & c \end{bmatrix}$$

and

$$[\zeta]^T = \begin{bmatrix} c & -s \\ s & c \end{bmatrix}$$

and the inverses of these can be obtained from (1.31), as follows:

$$[\zeta]^{-1} = \frac{\begin{bmatrix} c & -s \\ s & c \end{bmatrix}}{c^2 + s^2} = \begin{bmatrix} c & -s \\ s & c \end{bmatrix},$$

i.e. for *orthogonal matrices*

$$[\zeta]^{-1} = [\zeta]^T$$

1.5.5 Diagonal matrices

The inverse of a diagonal matrix is simply obtained by finding the reciprocal of each element on the leading diagonal, as shown in (1.64):

$$\begin{bmatrix} A_{11} & 0 & 0 & 0 & 0 \\ 0 & A_{22} & 0 & 0 & 0 \\ 0 & 0 & A_{33} & 0 & 0 \\ 0 & 0 & 0 & A_{44} & 0 \\ 0 & 0 & 0 & 0 & A_{55} \end{bmatrix}^{-1} = \begin{bmatrix} 1/A_{11} & 0 & 0 & 0 & 0 \\ 0 & 1/A_{22} & 0 & 0 & 0 \\ 0 & 0 & 1/A_{33} & 0 & 0 \\ 0 & 0 & 0 & 1/A_{44} & 0 \\ 0 & 0 & 0 & 0 & 1/A_{55} \end{bmatrix}.$$

$$(1.64)$$

Some rules involving inverted matrices

$$([A][B])^{-1} = [B]^{-1}[A]^{-1}$$

$$[A^T]^{-1} = [A]^{-1T}.$$

If

$$[A] = \begin{bmatrix} B & 0 \\ \hline 0 & I \end{bmatrix}$$

then

$$[A]^{-1} = \begin{bmatrix} B^{-1} & 0 \\ \hline 0 & I \end{bmatrix}.$$

Similarly, if

$$[A] = \begin{bmatrix} B & 0 \\ \hline 0 & C \end{bmatrix}$$

then

$$[A]^{-1} = \begin{bmatrix} B^{-1} & 0 \\ \hline 0 & C^{-1} \end{bmatrix}.$$

1.6 SOLUTION OF SIMULTANEOUS EQUATIONS

The solution of simultaneous equations is a frequently recurring problem in science and engineering. Basically, there are two types of simultaneous equation: *homogeneous* (1.65) and *non-homogeneous* (1.66). From (1.65) and (1.66), it can be seen that the difference between the two is that, for homogeneous equations, the terms on the right are zero.

$$
\begin{aligned}
A_{11}x_1 + A_{12}x_2 + \ldots\ldots + A_{1n}x_n &= 0 \\
A_{21}x_1 + A_{22}x_2 + \ldots\ldots + A_{2n}x_n &= 0 \\
&\vdots \\
A_{n1}x_1 + A_{n2}x_2 + \ldots\ldots + A_{nn}x_n &= 0
\end{aligned}
\tag{1.65}
$$

$$A_{11}x_1 + A_{12}x_2 + \ldots\ldots + A_{1n}x_n = C_1$$
$$A_{21}x_1 + A_{22}x_2 + \ldots\ldots + A_{2n}x_n = C_2$$

$$A_{n1}x_1 + A_{n2}x_2 + \ldots\ldots + A_{nn}x_n = C_n. \tag{1.66}$$

The solution of non-homogeneous equations will be covered in this section, whilst the solution of homogeneous equations, which are much more difficult, will be considered in section 1.7.

Equations (1.66) can be put in matrix form, as shown in (1.67):

$$[A]\{x\} = \{C\} \tag{1.67}$$

where

$[A]$ and $\{C\}$ are known

and

$\{x\}$ is a vector of the unknowns.

One method of solving (1.67) is to invert $[A]$, and premultiply both sides by $[A]^{-1}$, as shown in (1.68):

$$[A]^{-1}[A]\{x\} = [A]^{-1}\{C\} \tag{1.68}$$

which results in a solution for $\{x\}$ as follows,

$$\{x\} = [A]^{-1}\{C\}. \tag{1.69}$$

Equation (1.69) is a very inefficient method of solving simultaneous equations, as operations on $[A]$ have to take place, both above and below the leading diagonal.

There are a number of much more efficient methods of solving simultaneous equations, and some of these are described below.

1.6.1 Cramer's rule

This is one of the most common methods of solving small sets of simultaneous equations. Cramer's rule, which employs determinants, is given in (1.70).

$$
x_i = \frac{\begin{vmatrix} A_{11} & A_{12} & \cdots & A_{1,i-1} & C_1 & A_{1,i+1} & \cdots & A_{1n} \\ A_{21} & A_{22} & \cdots & A_{2,i-1} & C_2 & A_{2,i+1} & \cdots & A_{2n} \\ \cdot & \cdot & & \cdot & \cdots & & & \cdot \\ \cdot & \cdot & & \cdot & \cdot & \cdot & & \cdot \\ \cdot & \cdot & & \cdot & \cdot & \cdot & & \cdot \\ A_{n1} & A_{n2} & & A_{n,i-1} & C_n & A_{n,i+1} & \cdots & A_{nn} \end{vmatrix}}{|A|}
$$

$$
\text{ith column} \downarrow \tag{1.70}
$$

where,

$\qquad x_i = i$th unknown.

From (1.70), it can be seen that to obtain the ith unknown, the ith column of $|A|$ is replaced by the vector $\{C\}$ and then by expanding the two determinants, x_i can be determined.

1.6.2 Gauss' elimination method

Consider the set of equations:

$$
\begin{aligned}
A_{11}x_1 + A_{12}x_2 + \ldots + A_{1n}x_n &= C_1 \\
A_{21}x_1 + A_{22}x_2 + \ldots + A_{2n}x_n &= C_2 \\
&\vdots \\
A_{n1}x_1 + A_{n2}x_2 + \ldots + A_{nn}x_n &= C_n
\end{aligned} \tag{1.71}
$$

From the first equation of (1.71):

$$
x_1 = \frac{1}{A_{11}} \left(C_1 - \sum_{i=2}^{n} A_{1i}x_i \right) \tag{1.72}
$$

Substituting this value of x_1 into the second and other equations (down to the nth equation), a set of $n-1$ equations are obtained as shown in (1.73).

$$
\begin{aligned}
A^1_{22}x_2 + A^1_{23}x_3 + \ldots + A^1_{2n}x_n &= C^1_2 \\
A^1_{32}x_2 + A^1_{33}x_3 + \ldots + A^1_{3n}x_n &= C^1_3 \\
&\vdots \\
A^1_{n2}x_2 + A^1_{n3}x_3 + \ldots + A^1_{nn}x_n &= C^1_n .
\end{aligned} \tag{1.73}
$$

The above process can be repeated by eliminating x_2, x_3 etc., until two equations are remaining, as in (1.74):

$$B_{n-1,n-1}x_{n-1} + B_{n-1,n}x_n = k_{n-1}$$
$$B_{n,n-1}x_{n-1} + B_{nn}x_n = k_n \tag{1.74}$$

Finally, from (1.74),

$$B^1_{nn}x_n = k^1_n \tag{1.75}$$

Calculation of x_n can be made from (1.75); and from (1.73), (1.74), etc., the other unknowns can be determined.

1.6.3 Triangulation method

Consider the set of equations (1.71). Divide the first equation by A_{11} to give the first equation of (1.76). Multiply the first equation of (1.76) by A_{21} and take the resulting equation away from the second equation of (1.71) to give the second equation of (1.76). Now multiply the first equation of (1.76) by A_{31} and take the resulting equation away from the third equation of (1.71) to give the third equation of (1.76). If this process is repeated to the nth equation, then the following set of simultaneous equations is obtained:

$$x_1 + A^1_{12}x_2 + \quad . \quad . \quad . \quad + A^1_{1n}x_n = C^1_1$$
$$A^1_{22}x_2 + \quad . \quad . \quad . \quad + A^1_{2n}x_n = C^1_1$$
$$\vdots \qquad\qquad\qquad \vdots \qquad\quad \vdots$$
$$A^1_{n2}x_2 + \quad . \quad . \quad . \quad + A^1_{nn}x_n = C^1_n \tag{1.76}$$

Divide the second equation of (1.76) by A^1_{22} to give the second equation of (1.77). Multiply this equation by A^1_{32}, A^1_{42}, etc., and take these equations away from the third, fourth, etc., equations of (1.76) respectively to eliminate x_2 from the latter. If the process is continued, the equations will eventually be in triangular form as shown in (1.77):

$$x_1 + A_{12}{}^1x_2 + \quad . \quad . \quad . \quad . \quad . \quad . \quad + A_{1n}{}^1x_n = C^1_1$$
$$x_2 + \quad . \quad . \quad . \quad . \quad . \quad . \quad + B_{2n}x_n = k_2$$
$$\vdots$$
$$x_{n-1} + B_{n-1,n}x_n = k_{n-1}$$
$$x_n = k_n \tag{1.77}$$

Once x_n is known, the other unknowns can be obtained by back-substitution in the order $x_{n-1}, x_{n-2}, x_{n-3} \ldots \ldots x_1$.

Example 1.8

To illustrate the method, the simultaneous equations of (1.78) will be considered.

$$2x_1 + 4x_2 + 3x_3 = 2$$
$$x_1 - 2x_2 \qquad = 3$$
$$-x_1 - 4x_2 + 5x_3 = 1.$$ (1.78)

Divide the first of the above equations by A_{11} (the first pivot) to give the following set of simultaneous equations:

$$x_1 + 2x_2 + 1.5x_3 = 1$$
$$x_1 - 2x_2 \qquad = 3$$
$$-x_1 - 4x_2 + 5x_3 = 1.$$ (1.79)

To eliminate x_1 from the 2nd row of (1.79), multiply the 1st row of (1.79) by A_{21}, which is one in this case, and take away the result from row 2 of (1.79).

$$x_1 + 2x_2 + 1.5x_3 = 1$$
$$-4x_2 - 1.5x_3 = 2$$
$$-x_1 - 4x_2 + 5x_3 = 1.$$ (1.80)

To eliminate $(-x_1)$ from (1.80), multiply the 1st row of (1.80) by A_{31}, which is minus one in this case, and take away the result from row 3 of (1.80).

$$x_1 + 2x_2 + 1.5x_3 = 1$$
$$-4x_2 - 1.5x_3 = 2$$ (1.81)
$$-2x_2 + 6.5x_3 = 2.$$

Divide the 2nd row of (1.81) by A_{22}, which is the second pivot, to give the second row of (1.82).

To eliminate x_2 from the 3rd row of (1.81), multiply the 2nd row of (1.82) by A_{32} of (1.81), and take away the result from the 3rd row of (1.81).

$$x_1 + 2x_2 + 1.5x_3 = 1$$
$$x_2 + 0.375x_3 = -0.5$$
$$7.25x_3 = 1.$$ (1.82)

Equations (1.82) can be seen to be of the triangular form shown by (1.77), and from (1.82):

$$x_3 = 0.1379$$

Hence, by back-substitution in the 2nd and 3rd rows of (1.82), x_2 and x_1 can be determined.

This method of solving simultaneous equations is very powerful and is in many respects similar to Chio's method for evaluating determinants.

Table 1.3 contains a program for solving simultaneous equations through triangulation, and this can be seen to be an extension of the program of Table 1.2. The program breaks down if a zero is met on the leading diagonal, but this can be overcome by interchanging two columns or rows, and interchanging the corresponding unknowns prior to outputting their values.

Table 1.3. Program for solving simultaneous equations.

```
100 REM SOLUTION OF SIMULTANEOUS EQUATIONS
110 INPUT"TYPE IN THE NUMBER OF SIMULTANEOUS EQUATIONS";N
120 DIM A(N,N),C(N),X(N)
130 PRINT"TYPE IN THE MATRIX (A)"
140 FOR I=1 TO N
150 FOR J=1 TO N
160 PRINT"TYPE IN A(";I;",";J;")"
170 INPUT A(I,J)
180 NEXT J,I
190 PRINT"TYPE IN THE VECTOR (C)"
200 FOR I=1 TO N
210 PRINT"C(";I;")=";
220 INPUTC(I)
230 NEXT I
250 FOR I=1TO N-1
260 PI=A(I,I)
270 FOR J=I TO N
280 A(I,J)=A(I,J)/PI
290 NEXT J
300 C(I)=C(I)/PI
310 FOR K=I TO N-1
320 CN=A(K+1,I)
330 FOR J=I TO N
340 A(K+1,J)=A(K+1,J)-A(I,J)*CN
350 NEXT J
360 C(K+1)=C(K+1)-C(I)*CN
370 NEXTK
380 NEXT I
390 X(N)=C(N)/A(N,N)
400 FORI=N-1 TO 1 STEP -1
410 FORJ=I+1 TO N
420 X(I)=X(I)-X(J)*A(I,J)
430 NEXT J
440 X(I)=X(I)+C(I)
450 NEXT I
460 PRINT"THE RESULTS ARE AS FOLLOWS :-"
470 FOR I=1 TO N
480 PRINTX(I)
490 NEXT I
500 END
```

1.6.4 Gauss-Seidel method

Consider the set of simultaneous equations (1.71). Solving for x_1 from the first of these equations gives

$$x_1 = \frac{1}{A_{11}}(C_1 - A_{12}x_2 - A_{13}x_3 \ . \ . \ . \ . \ . \ - A_{1n}x_n). \tag{1.83}$$

Setting x_2, x_3 to x_n to zero, the first estimate for x_1 can be obtained from (1.83).

From the second equation of (1.71),

$$x_2 = \frac{1}{A_{22}} (C_2 - A_{21}x_1 - A_{23}x_3 \ \ldots \ - A_{2n}x_n) \qquad (1.84)$$

Substituting the first estimate of x_1 into (1.84) and setting x_3 to x_n to zero, the first estimate of x_2 can be determined.

This process can be repeated so that

$$x_n = \frac{1}{A_{nn}} (C_n - A_{n1}x_1 - A_{n2}x_2 \ \ldots \ - A_{n,\,n-1}x_{n-1}) \qquad (1.85)$$

The first estimates of x_2 to x_n must now be substituted in (1.83) to calculate the second estimate of x_1, and this, together with the first estimates of x_3 to x_n, must be substituted into (1.84) to determine the second estimate of x_2. This procedure should be continued for the second estimates of all other values of x_i, and comparison must then be made of the second estimates with those of the first. If these do not show convergence, then further iterations must be carried out. There are of course some sets of simultaneous equations whose solution by this method is unsuitable, and equations which are not diagonally dominant are an example of this.

Equations which are particularly suitable for solution by this method, however, are those which are dominated by their diagonal terms, such as in structural analysis.

1.6.5 Solution of banded type equations

In many practical problems, large sets of simultaneous equations with all their non-zero elements lying in a narrow band about the main diagonal (1.10) have to be solved. There are several methods of solving these, and the method by Wilson [24], based on Gaussian elimination, is given here.

The solution by this method is to arrange the equations such that a set of tridiagonal equations are obtained, the elements of these being themselves small matrices as in (1.86).

$$
\begin{aligned}
[A_{11}]\{x_1\} + [A_{12}]\{x_2\} \qquad\qquad\qquad\quad &= \{C_1\} \\
[A_{21}]\{x_1\} + [A_{22}]\{x_2\} + [A_{23}]\{x_3\} \qquad &= \{C_2\} \\
[A_{32}]\{x_2\} + [A_{33}]\{x_3\} + [A_{34}]\{x_4\} &= \{C_3\}
\end{aligned}
$$

$$\cdot \ \cdot \ \cdot \ \cdot \ \cdot \ \cdot$$

$$[A_{n,\,n-1}]\{x_{n-1}\} + [A_{nn}]\{x_n\} = \{C_n\}. \qquad (1.86)$$

Eliminating equations (1.86) in a manner similar to that of the triangulation method, the following are obtained:

$$\{x_n\} \ = [a_{nn}]^{-1}\{k_n\}$$

$$\{x_{n-1}\} = [a_{n-1,\,n-1}]^{-1}\{k_{n-1}\} - [a_{n-1,\,n-1}]^{-1}[A_{n-1,\,n}]\{x_n\}$$

$$\vdots \qquad\qquad\qquad \vdots \qquad\qquad\qquad \vdots$$

$$\{x_1\} \ \ = [a_{11}]^{-1}\{k_1\} - [a_{11}]^{-1}[A_{12}]\{x_2\},$$

where

$$[a_{11}] = [A_{11}]$$
$$[a_{22}] = [A_{22}] - [A_{21}][a_{11}]^{-1}[A_{12}]$$

$$\cdot \qquad \cdot \qquad \cdot$$
$$\cdot \qquad \cdot \qquad \cdot$$

$$[a_{nn}] = [A_{nn}] - [A_{n,\,n-1}][a_{n-1,\,n-1}]^{-1}[A_{n-1,\,n}]$$

and

$$\{k_1\} = \{C_1\}$$
$$\{k_2\} = \{C_2\} - [A_{21}][a_{11}]^{-1}\{k_1\}$$
$$\{k_n\} = \{C_n\} - [A_{n,\,n-1}][a_{n-1,\,n-1}]^{-1}\{k_{n-1}\}.$$

1.6.6 Alternative triangulation method for symmetrical banded simultaneous equations

An alternative method to that of Wilson's, which is particularly suitable for symmetrical matrices, is to consider only half the band, and to rotate this part of the matrix so that it is of the rectangular form of (1.87), where N is the number of simultaneous equations and NHW the half bandwidth.

$$\begin{bmatrix} A_{11} & & \\ A_{22} & & \\ A_{33} & & \\ \cdot & & \\ \cdot & & \\ A_{nn} & & \text{zero} \end{bmatrix} \Bigg\updownarrow N$$

$$\longleftarrow \text{NHW} \longrightarrow \qquad \cdot \tag{1.87}$$

From (1.87), it can be seen that the matrix is the upper half of the band of (1.10), plus some zeros, and also that the leading diagonal of (1.10) lies in the first column of (1.87).

Solution of the simultaneous equations is carried out by the triangulation process described in Section 1.6.3, except that the property of symmetry is used to save space.

Table 1.4 contains a computer program in BASIC, which solves symmetrical banded simultaneous equations, where the equations are stored as in equation (1.87).

The advantages of this method of solution can readily be seen, both from the point of view of saving space and time, and also because of the increased precision obtained.

Table 1.4. Program for solution of banded equations of symmetrical form.

```
100 REM SOLVEBAND
110 PRINT "VALUE OF N=";
120 INPUT N
130 PRINT "SIZE OF HALF BANDWIDTH=";
140 INPUT NW
150 NT=N+NW
160 DIM A(NT,NW),C(NT)
170 PRINT"TYPE IN THE HALF-BANDED FORM OF (A)"
180 FOR II=1 TO N
190 FOR JJ=1 TO NW
200 PRINT"A(";II;",";JJ;")=";
210 INPUT A(II,JJ)
220 NEXT JJ
230 NEXT II
240 PRINT "FEED IN VECTOR ON RHS (C)"
250 FOR II=1 TO N
260 PRINT "POSITION";II
270 INPUT C(II)
280 NEXT II
290 FOR II=1 TO N
300 IK=II
310 FOR JJ=2 TO NW
320 IK=IK+1
330 CN=A(II,JJ)/A(II,1)
340 JK=0
350 FOR KK=JJ TO NW
360 JK=JK+1
370 A(IK,JK)=A(IK,JK)-CN*A(II,KK)
380 NEXT KK
390 A(II,JJ)=CN
400 C(IK)=C(IK)-CN*C(II)
410 NEXT JJ
420 C(II)=C(II)/A(II,1)
430 NEXT II
440 II=N
450 II=II-1
460 IF II<=0 THEN 530
470 JJ=II
480 FOR KK=2 TO NW
490 JJ=JJ+1
500 C(II)=C(II)-A(II,KK)*C(JJ)
510 NEXT KK
520 GOTO 450
530 PRINT"THE VECTOR ON THE LHS (X) IS:-"
540 FOR II=1 TO N
550 PRINT C(II)
560 NEXT II
570 END
```

1.7 SOLUTION OF EIGENVALUES AND EIGENVECTORS

Eigenvalues or the roots of a polynomial are also known as *characteristic values*.

Equations such as (1.88) are frequently met in science and engineering, (e.g. vibrations, oscillations, and buckling), and they constitute an eigenvalue problem.

$$[A]\{x\} = \{Y\}. \tag{1.88}$$

If λ = eigenvalues, then (1.88) becomes

$$[A]\{x\} = \lambda\{x\}. \tag{1.89}$$

Equation (1.89) is known as the *eigenvector equation*, and it can be represented by the homogeneous equations of (1.90) and (1.65):

$$([A] - \lambda[I])\{x\} = \{0\} \tag{1.90}$$

where the solutions for $\{x\}$ are known as *eigenvectors*.

In most practical problems the condition that $\{x\}$ is null is not of interest, hence the following should be satisfied.

$$\left|[A] - \lambda[I]\right| = \begin{vmatrix} A_{11} - \lambda & A_{12} & . & . & . & A_{1n} \\ A_{21} & A_{22} - \lambda & . & . & . & A_{2n} \\ . & . & . & & & \\ . & . & . & & & \\ . & . & . & . & & \\ A_{n1} & A_{n2} & . & . & . & A_{nn} - \lambda \end{vmatrix} = 0 \tag{1.91}$$

Equation (1.91) is known as the *characteristic* equation.

Example 1.9
Determine the eigenvalues and eigenvectors for the matrix of equation (1.92):

$$[A] = \begin{bmatrix} 4 & 2 \\ 1 & 8 \end{bmatrix}. \tag{1.92}$$

To determine the eigenvalues, (1.92) must be written in the form shown in (1.93):

$$\begin{vmatrix} (4 - \lambda) & 2 \\ 1 & (8 - \lambda) \end{vmatrix} = 0 \tag{1.93}$$

where the eigenvalues are given by λ.

Expansion of (1.93) gives (1.94):

$$(4 - \lambda)(8 - \lambda) - (2 \times 1) = 0$$

or

$$\lambda^2 - 12\lambda + 30 = 0. \tag{1.94}$$

The eigenvalues or roots of (1.94) are

$$\underline{\lambda_1 = 8.449}$$

and

$$\underline{\lambda_2 = 3.551}.$$

λ_1 is known as the *dominant eigenvalue*, as it is the largest in magnitude, and λ_2 is known as the *sub-dominant* eigenvalue, as it is the second largest in magnitude.

To determine the eigenvectors, substitute λ_1 and λ_2 into (1.90):

$$\begin{bmatrix} (4 - \lambda_1) & 2 \\ 1 & (8 - \lambda_2) \end{bmatrix} \begin{Bmatrix} x_1 \\ x_2 \end{Bmatrix} = \{0\}$$

which gives an eigenvector *corresponding to* λ_1 of

$$\begin{Bmatrix} x_1 \\ x_2 \end{Bmatrix} = \begin{Bmatrix} 0.45 \\ 1 \end{Bmatrix}$$

and an eigenvector *corresponding to* λ_2 of

$$\begin{Bmatrix} 1 \\ -0.225 \end{Bmatrix}.$$

The matrix of (1.92) is said to be *positive definite* as both its roots are positive.

From the above it can be seen that the calculation of eigenvalues by the expansion of a determinant is limited only to small matrices, as an nth order matrix would involve a polynomial of the nth degree. There are a number of methods of calculating eigenvalues for larger matrices, but only one will be given in the present text.

1.7.1 The power method

This method is often used to calculate the largest (dominant) eigenvalue of the equation (1.89). The process is as follows.

Assume any arbitrary values for $\{x\}$ in (1.88) and then calculate $\{Y\}$ from the expression

$$\{Y\} = [A]\{x\}. \tag{1.95}$$

From (1.89) and (1.95), an estimate for λ is

$$\lambda = \sum_1^n Y_i \bigg/ \sum_1^n x_i,$$

and the new value of $\{x\}$ is given by

$$\{x\} = \text{constant } \{Y\}$$

where the constant has been introduced to keep the size of $\{x\}$ down to a suitable value. Substituting this value of $\{x\}$ back into (1.95) will show a convergence for λ. This process is, however, very slow; but for positive eigenvalues, the following procedure overcomes this difficulty:

$$\lambda = \sqrt{(\{Y\}^{\mathrm{T}}\{Y\})} \tag{1.95}$$

and from (1.88) and (1.89)

$$\{x\} = \frac{\{Y\}}{\lambda}. \tag{1.97}$$

Now substitute $\{x\}$ from (1.97) into (1.95) and repeat the process until convergence is obtained for λ.

1.7.2 Orthogonality relations for real symmetric matrices

If $x_1, x_2, x_3, \ldots, x_n$ are the elements of the eigenvector corresponding to λ, and $x_1{}^1, x_2{}^1, x_3{}^1, \ldots, x_n{}^1$ are the elements of the eigenvector corresponding to λ^1, the relation between the two vectors is

$$x_1 x_1{}^1 + x_2 x_2{}^1 + x_3 x_3{}^1 + \ldots + x_n x_n{}^1 = 0. \tag{1.98}$$

1.7.3 Calculation of other eigenvalues and eigenvectors

The property of equation (1.98) can be used to calculate other eigenvalues and eigenvectors for real symmetric matrices as follows.

Assuming that the largest eigenvalue has been obtained, the next highest value can be determined by a similar process.

$$\begin{bmatrix} A_{11} & A_{12} & \cdots & \cdots & A_{1n} \\ A_{21} & A_{22} & \cdots & \cdots & A_{2n} \\ \cdot & \cdot & & & \\ \cdot & \cdot & & & \\ \cdot & \cdot & & & \\ A_{n1} & A_{n2} & \cdots & \cdots & A_{nn} \end{bmatrix} \begin{Bmatrix} x_1{}^1 \\ x_2{}^1 \\ \cdot \\ \cdot \\ \cdot \\ x_n{}^1 \end{Bmatrix} = \lambda^1 \begin{Bmatrix} x_1{}^1 \\ x_2{}^1 \\ \cdot \\ \cdot \\ \cdot \\ x_n{}^1 \end{Bmatrix}. \tag{1.99}$$

Solving (1.98) for $x_n{}^1$,

$$x_n{}^1 = \frac{-x_1 x_1{}^1}{x_n} \frac{-x_2 x_2{}^1}{x_n} \frac{-x_3 x_3{}^1}{x_n} \cdots \cdots \frac{-x_{n-1} x_{n-1}{}^1}{x_n}. \tag{1.100}$$

If (1.99) is now written in the form of simultaneous equations and (1.100) is substituted into these, n equations are obtained with $n-1$ unknowns. As the last equation is redundant, it can be dropped to give

$$
\begin{bmatrix}
B_{11} & B_{12} & \cdot & \cdot & \cdot & \cdot & \cdot & B_{1,\,n-1} \\
B_{21} & B_{22} & \cdot & \cdot & \cdot & \cdot & \cdot & B_{2,\,n-1} \\
\cdot & \cdot & & & & & & \cdot \\
\cdot & \cdot & & & & & & \cdot \\
\cdot & \cdot & & & & & & \cdot \\
\cdot & \cdot & & & & & & \cdot \\
B_{n-1,\,1} & B_{n-1,\,2} & \cdot & \cdot & \cdot & \cdot & \cdot & B_{n-1,\,n-1}
\end{bmatrix}
\begin{Bmatrix}
x_1{}^1 \\ x_2{}^1 \\ \cdot \\ \cdot \\ \cdot \\ \cdot \\ x_{n-1}{}^1
\end{Bmatrix}
= \lambda^1
\begin{Bmatrix}
x_1{}^1 \\ x_2{}^1 \\ \cdot \\ \cdot \\ \cdot \\ \cdot \\ x_{n-1}{}^1
\end{Bmatrix}.
$$

$$(1.101)$$

Equation (1.101) can now be solved by the power method, and by back-substitution into (1.100) the corresponding eigenvector can be determined.

Equations of the type shown in (1.102) are often met in practice.

$$[A^1]\{x\} - \lambda[B]\{x\} = \{0\}. \tag{1.102}$$

These can be reduced to the form of (1.89) by premultiplying both sides by $[B]^{-1}$, so that

$$[B]^{-1}[A^1]\{x\} - \lambda\{x\} = 0$$

or

$$[A]\{x\} = \lambda\{x\}.$$

1.7.4 Rayleigh quotient

This is a useful method for finding the bounds of eigenvalues for positive definite symmetric matrices.

The Rayleigh quotient 'r' is defined as

$$r = \frac{\lfloor x \rfloor [A]\{x\}}{\lfloor x \rfloor \{x\}}$$

where

$$\lambda_1 \leq r \geq \lambda_n.$$

1.8 INTEGRATION AND DIFFERENTIATION OF MATRICES

If

$$
[A] = \begin{bmatrix}
4x & -5x^3 & 6 \\
7x^2 & 8x & -9x^4 \\
2x^3 & -7 & 5x^2
\end{bmatrix},
$$

then the derivative of $[A]$ with respect to x is obtained by differentiating each element of $[A]$, as follows:

$$\frac{d[A]}{dx} = \begin{bmatrix} 4 & -15x^2 & 0 \\ 14x & 8 & -36x^3 \\ 6x^2 & 0 & 10x \end{bmatrix},$$

Similarly, if the integral of $[A]$ is required between the limits from 0 to 1, each element is integrated in turn, as follows:

$$\int_0^1 [A]dx = \begin{bmatrix} \int_0^1 4x\,dx & \int_0^1 -5x^3\,dx & \int_0^1 6dx \\ \int_0^1 7x^2\,dx & \int_0^1 8x\,dx & \int_0^1 -9x^4\,dx \\ \int_0^1 2x^3\,dx & \int_0^1 -7dx & \int_0^1 5x^2\,dx \end{bmatrix}$$

$$= \begin{bmatrix} 2 & -\frac{5}{4} & 6 \\ \frac{7}{3} & 4 & -\frac{9}{5} \\ \frac{1}{2} & -7 & \frac{5}{3} \end{bmatrix},$$

EXAMPLES FOR PRACTICE

If

$$[A] = \begin{bmatrix} 2 & 1 \\ 3 & 4 \end{bmatrix} \qquad [B] = \begin{bmatrix} 4 & -1 \\ 0 & -4 \end{bmatrix}$$

$$[C] = \begin{bmatrix} 2 & -1 & 3 \\ 4 & -2 & 1 \end{bmatrix}$$

$$[D] = \begin{bmatrix} 4 & 1 & 2 \\ -2 & 6 & -3 \\ 0 & 3 & 8 \end{bmatrix}$$

$$[E] = \begin{bmatrix} 6 & 2 & -1 \\ -1 & 4 & 0 \\ -2 & 3 & 10 \end{bmatrix}$$

$$[F] = \begin{bmatrix} 2 & -1 & 0 & 0 \\ -1 & 2 & -1 & 0 \\ 0 & -1 & 2 & -1 \\ 0 & 0 & -1 & 2 \end{bmatrix}$$

$$[G] = \begin{bmatrix} 8 & -2 & -1 & 3 \\ 1 & 6 & 2 & -1 \\ 2 & -3 & 6 & 4 \\ 0 & -4 & 5 & 10 \end{bmatrix}$$

$$\lfloor H \rfloor = \begin{bmatrix} 1 & 2 & 3 & 4 \end{bmatrix}$$

find:

(1)   transpose of $[A]$

(2)   transpose of $[B]$

(3)   transpose of $[C]$

(4)   transpose of $[D]$

(5)   transpose of $[G]$

(6) $[A] + [B]$

(7) $[A] - [B]$

(8) $[D] + [E]$

(9) $[D] - [E]$

(10) $[A][B]$

(11) $[B][A]$

(12) $[A][C]$

(13) $[B][C]$

(14) $[D][E]$

(15) $[E][D]$

(16) $[F][G]$

(17) $\lfloor H \rfloor \{H\}$

(18) $\{H\} \lfloor H \rfloor$

(19) determinant of $[A]$

(20) determinant of $[B]$

(21) determinant of $[D]$

(22) determinant of $[E]$

(23) determinant of $[F]$

(24) inverse of $[A]$

(25) inverse of $[B]$

(26) inverse of $[D)$

(27) inverse of $[E]$

(28) inverse of $[F]$

(29) eigenvalues of $[A]$

(30) eigenvalues of $[B]$

(31) eigenvalues of $[E]$

(32) eigenvalues of $[F]$

Solve the following sets of simultaneous equations :

(33) $2x_1 - x_2 + 3x_3 = 1$

$-x_1 + 2x_2 + x_3 = 2$

$3x_1 + 3x_2 - x_3 = 3$

(34) $6x_1 - 2x_2 + x_3 = 4$

$x_1 - 2x_2 - x_3 = 1$

$2x_1 + x_2 + x_3 = 5$

(35) $2x_1 - x_2 \qquad\qquad = 1$

$-x_1 + 2x_2 - x_3 \qquad = 2$

$- x_2 + 2x_3 - x_4 = 3$

$- x_3 + 2x_4 = 4.$

Answers

(1) $\begin{bmatrix} 2 & 3 \\ 1 & 4 \end{bmatrix}$

(2) $\begin{bmatrix} 4 & 0 \\ -1 & -4 \end{bmatrix}$

(3) $\begin{bmatrix} 2 & 4 \\ -1 & -2 \\ 3 & 1 \end{bmatrix}$

(4) $\begin{bmatrix} 4 & -2 & 0 \\ 1 & 6 & 3 \\ 2 & 3 & 8 \end{bmatrix}$

(5) $\begin{bmatrix} 8 & 1 & 2 & 0 \\ -2 & 6 & -3 & -4 \\ -1 & 2 & 6 & 5 \\ 3 & -1 & 4 & 10 \end{bmatrix}$

(6) $\begin{bmatrix} 6 & 0 \\ 3 & 0 \end{bmatrix}$

(7) $\begin{bmatrix} -2 & 2 \\ 3 & 8 \end{bmatrix}$

(8) $\begin{bmatrix} 10 & 3 & 1 \\ -3 & 10 & -3 \\ -2 & 6 & 18 \end{bmatrix}$

(9) $\begin{bmatrix} -2 & -1 & 3 \\ -1 & 2 & -3 \\ 2 & 0 & -2 \end{bmatrix}$

(10) $\begin{bmatrix} 8 & -16 \\ 12 & -19 \end{bmatrix}$

(11) $\begin{bmatrix} 5 & 0 \\ -12 & -16 \end{bmatrix}$

(12) $\begin{bmatrix} 8 & -4 & 7 \\ 22 & -11 & 13 \end{bmatrix}$

(13) $\begin{bmatrix} 4 & -2 & 11 \\ -16 & 8 & -4 \end{bmatrix}$

(14) $\begin{bmatrix} 19 & 18 & 16 \\ -12 & 11 & -28 \\ -19 & 36 & 80 \end{bmatrix}$

(15) $\begin{bmatrix} 20 & 15 & -2 \\ -12 & 23 & -14 \\ -14 & 46 & 67 \end{bmatrix}$

(16) $\begin{bmatrix} 15 & -10 & -4 & 7 \\ -8 & 17 & -1 & -9 \\ 3 & -8 & 5 & -1 \\ -2 & -5 & 4 & 16 \end{bmatrix}$

(17)   30

(18) $\begin{bmatrix} 1 & 2 & 3 & 4 \\ 2 & 4 & 6 & 8 \\ 3 & 6 & 9 & 12 \\ 4 & 8 & 12 & 16 \end{bmatrix}$

(19)   5

(20) -16

(21) 232

(22) 255

(23)   5

$$(24) \begin{bmatrix} 0.8 & -0.2 \\ -0.6 & 0.4 \end{bmatrix}$$

$$(25) \begin{bmatrix} 0.25 & -0.0625 \\ 0 & -0.25 \end{bmatrix}$$

$$(26) \begin{bmatrix} 0.2457 & -8.621\text{E-}3 & -0.0647 \\ 0.0690 & 0.1379 & 0.0345 \\ -0.0259 & -0.0517 & 0.1121 \end{bmatrix}$$

$$(27) \begin{bmatrix} 0.1567 & -0.0902 & 0.01569 \\ 0.0392 & 0.2275 & 3.922\text{E-}3 \\ 0.0196 & -0.0863 & 0.1020 \end{bmatrix}$$

$$(28) \begin{bmatrix} 0.8 & 0.6 & 0.4 & 0.2 \\ 0.6 & 1.2 & 0.8 & 0.4 \\ 0.4 & 0.8 & 1.2 & 0.6 \\ 0.2 & 0.4 & 0.6 & 0.8 \end{bmatrix}$$

(29) 5; $\lfloor 0.333 \quad 1 \rfloor$
    1; $\lfloor 1 \quad -1 \rfloor$

(30) -4; $[1 \quad 1]$
    4; $[1 \quad 0]$

(31) 10.51; $[-0.207 \quad 0.031 \quad 1]$
    6.83; $[1 \quad 0.16 \quad 0.49]$
    3.58; $[-0.77 \quad 1 \quad -0.71]$

(32) 2.62; $[1 \quad -0.62 \quad -0.62 \quad 1]$
    3.62; $[-0.63 \quad 1 \quad -0.99 \quad 0.61]$
    1.38; $[1 \quad 0.62 \quad -0.62 \quad -1]$
    0.38; $[0.63 \quad 1 \quad 0.99 \quad 0.61]$

(33) $x_1 = 0.256$; $x_2 = 0.897$; $x_3 = 0.462$

(34) $x_1 = 3.8$; $x_2 = 5.4$; $x_3 = -8$

(35) $x_1 = 4$; $x_2 = 7$; $x_3 = 8$; $x_4 = 6$.

Chapter 2

Basic structural concepts and energy theorems

The theoretical analysis of structures is based on either an analytical solution or a numerical one, or some combination of the two.

The approach in this chapter is to consider stiffness and flexibility of simple elastic springs, and also to introduce the concepts of elemental and system stiffness matrices. Later, these terms are related to some of the major energy theorems which are popular in structural mechanics.

2.1 STIFFNESS AND FLEXIBILITY

Consider a uniform elastic spring subjected to a load P, as shown in Fig. 2.1. If

k = stiffness of spring
= load required to give the spring a unit displacement

and

u = displacement of spring due to a load P,

then

$$P = ku \qquad (2.1)$$

or

$$u = \frac{1}{k}P$$

$$= fP \qquad (2.2)$$

Fig. 2.1 – Uniform elastic spring.

where f = flexibility of spring (displacement/unit load).

Thus it can be seen that, once the stiffness of the spring is known, the displacement can be readily determined for any load P. It should also be noted that flexibility is the reciprocal or inverse of stiffness.

This elementary problem is in fact a very simplified description of how most structures behave. For in the elastic design of structures, most structures can be considered to be composed of thousands of interconnected "springs" of various stiffness in various directions and subjected to complex combinations of loads.

Thus, for a complex structure, it is necessary to consider the individual contribution of each "spring", and this is assisted by isolating the effects of each "spring" by a system of nodes.

For the practical problem, the "spring" is in fact called a finite element, and by using various combinations of finite elements, all interconnected at their boundaries, it is possible to describe a complex structure.

Although the structure might be of a complex shape, the elements are of a much simpler shape, so that it is a relatively easy matter to obtain element stiffness. Elements take many different forms, depending on the structure they are supposed to represent. For example, it is usual to use rod elements for pin-jointed trusses and "beam-column" elements for rigid-jointed frames, whilst triangular and quadrilateral elements are often used to represent plates and doubly-curved shells.

Nodal points or nodes are used to describe the individual elements and also the entire structure, and they are normally taken at convenient points in the structure, usually to simplify the element. For example, the nodes of a rod or beam element are usually at the ends of the element, whilst for triangles and quadrilaterals, they are usually taken at the corners of the element. Sometimes, however, additional "mid-side" nodes are used to define curved shapes.

2.2 STIFFNESS MATRIX

To analyse a more complicated structure, it will be necessary to extend the scalar expression of (2.1) into a matrix form. This is because the element behaviour must be described with respect to its nodal points.

Suppose the uniform elastic spring of Fig. 2.2 has nodal points 1 and 2 at its ends, and that the forces at these points are P_1 and P_2 with corresponding displacements u_1 and u_2.

Fig. 2.2 – Elemental spring.

$P_1 = k \times$ displacement of spring

$\quad = k(u_1 - u_2).$ $\hspace{4cm}$ (2.3)

From equilibrium considerations,

$P_2 = -P_1$

$\quad = k(u_2 - u_1).$ $\hspace{4cm}$ (2.4)

It is convenient to show (2.3) and (2.4) in matrix form as follows:

$$\begin{Bmatrix} P_1 \\ P_2 \end{Bmatrix} = \begin{bmatrix} k & -k \\ -k & k \end{bmatrix} \begin{Bmatrix} u_1 \\ u_2 \end{Bmatrix} \hspace{3cm} (2.5)$$

or

$$\{P_i\} = [k]\{u_i\} \hspace{3cm} (2.6)$$

where

$\{P_i\} =$ a vector of internal nodal forces $= \begin{Bmatrix} P_1 \\ P_2 \end{Bmatrix}$

$\{u_i\} =$ a vector of nodal displacements $= \begin{Bmatrix} u_1 \\ u_2 \end{Bmatrix}$

$[k] =$ the elemental stiffness matrix

$$= \begin{bmatrix} k & -k \\ -k & k \end{bmatrix} \hspace{3cm} (2.7)$$

Equation (2.5) cannot be solved, as $[k]$ is singular. In physical terms, this is because the spring is floating in space, and has not been constrained in any way.

Normally, as a structure can be regarded as being composed of many interconnected 'spring' elements, the following expression will be obtained on satisfying equilibrium and compatibility (see section 2.3).

$$\{q\} = [K]\{u_i\} \hspace{3cm} (2.8)$$

where

$\{q\} =$ a vector of external nodal forces $= \Sigma\{P_i\}$

$[K] =$ structural stiffness matrix

$\quad = \Sigma[k]$

$\{u_i\} =$ a vector of nodal displacements of the structure.

Expressions (2.6) and (2.8) can be seen to be of similar form.

It should be pointed out that (2.8) can only be solved if a sufficient number of constraints are introduced. The minimum number of required constraints will be those that prevent the structure from floating in space. This latter phenomenon is sometimes referred to as 'rigid body motion'.

2.3 CALCULATION OF STRUCTURAL STIFFNESS MATRIX

The calculation of the structural stiffness matrix is equivalent to assembling the entire structure.

Consider a structure composed of two springs in series, as shown in Fig. 2.3. As the two springs have different stiffnesses, it will be convenient to represent the structure with two elements. The first element will be spring (1), which will be bounded by the nodes, 1 and 2, and the second element will be the remaining spring bounded by the nodes 2 and 3. Let

Fig. 2.3 – Spring structure.

$$P_1 \quad = \text{internal force at node 1 in spring (1)}$$
$$P_2^{(1)} = \text{internal force at node 2 in spring (1)}$$
$$\quad\; = -P_1$$
$$P_2^{(2)} = \text{internal force at node 2 in spring (2)}$$
$$\quad\; = -P_3$$
$$P_3 = \text{internal force at node 3 in spring (2)}$$
$$Q_1 = \text{externally applied force at node 1} = \Sigma P_1$$
$$Q_2 = \text{externally applied force at node 2} = \Sigma \left[P_2^{(1)} + P_2^{(2)} \right]$$
$$Q_3 = \text{externally applied force at node 3} = \Sigma P_3$$
$$u_1 = \text{displacement at node 1}$$
$$u_2 = \text{displacement at node 2}$$
$$u_3 = \text{displacement at node 3.}$$

To obtain the overall relationship between forces and displacements, the effects of each nodal displacement on the internal forces 'P' will be considered, as if each were acting separately. Later, using the principle of superposition and by considering equilibrium, the equations relating all three forces and displacements will be obtained.

Case (a)

Let node 1 have a displacement u_1, and fix the spring at nodes 2 and 3. From (2.3),

$$P_1 = k_a(u_1 - 0)$$
$$= k_a u_1.$$

From (2.4),

$$P_2^{(1)} = -P_1 = k_a(0 - u_1)$$
$$= -k_a u_1.$$

As nodes 2 and 3 are fixed, $P_2^{(2)} = P_3 = 0$.

Case (b)

Let node 2 have a displacement of u_2, and fix the spring at nodes 1 and 3. From (2.3) and (2.4),

$$P_1 = k_a(0 - u_2) = -k_a u_2$$
$$P_3 = k_b(0 - u_2) = -k_b u_2.$$

From considerations of equilibrium,

$$P_2^{(1)} = -P_1 = k_a u_2$$
$$P_2^{(2)} = -P_3 = k_b u_2.$$

Case (c)

Let node 3 have a displacement of u_3, and fix the spring at nodes 1 and 2. From (2.3) and (2.4),

$$P_2^{(2)} = -k_b u_3$$
$$P_3 = k_b u_3.$$

As the spring is fixed at nodes 1 and 2, $P_1 = P_2^{(1)} = 0$.

To obtain the overall relationship, it will be necessary to consider the structure with all three displacements taking place together, and this condition can be met by superimposing Cases (a), (b), and (c), as follows.

Now,

$$Q_1 = \Sigma P_1 = k_a u_1 - k_a u_2$$
$$Q_2 = \Sigma[P_2^{(1)} + P_2^{(2)}] = -k_a u_1 + k_a u_2 + k_b u_2 - k_b u_3$$
$$= -k_a u_1 + (k_a + k_b)u_2 - k_b u_3 \qquad (2.9)$$
$$Q_3 = \Sigma P_3 = -k_b u_2 + k_b u_3.$$

Rewriting (2.9) in matrix form, the following is obtained:

$$\begin{Bmatrix} Q_1 \\ Q_2 \\ Q_3 \end{Bmatrix} = \begin{bmatrix} k_a & -k_a & 0 \\ -k_a & (k_a + k_b) & -k_b \\ 0 & -k_b & k_b \end{bmatrix} \begin{Bmatrix} u_1 \\ u_2 \\ u_3 \end{Bmatrix} \qquad (2.10)$$

or,

$$\{q\} = [K]\{u_i\}, \tag{2.11}$$

where

$$\{q\} = \begin{Bmatrix} Q_1 \\ Q_2 \\ Q_3 \end{Bmatrix} = \text{a vector of external nodal loads acting on the structure}$$

$$[K] = \begin{bmatrix} k_a & -k_a & 0 \\ -k_a & (k_a + k_b) & -k_b \\ 0 & -k_b & k_b \end{bmatrix} = \text{the system or structural stiffness matrix}$$

$$\{u_i\} = \begin{Bmatrix} u_1 \\ u_2 \\ u_3 \end{Bmatrix}.$$

2.3.1   Alternative method of forming $[K]$

The system stiffness matrix can be formed by directly superimposing the elemental stiffness matrices, as follows.

Element 1–2
From (2.5), the elemental stiffness matrix for element 1–2 is given by (2.12):

$$[k_{1-2}] = \begin{matrix} u_1 & u_2 \\ \begin{bmatrix} k_a & -k_a \\ -k_a & k_a \end{bmatrix} & \begin{matrix} u_1 \\ u_2 \end{matrix} \end{matrix}. \tag{2.12}$$

From (2.5), the elemental stiffness matrix for element 2–3 is given by (2.13):

$$[k_{2-3}] = \begin{matrix} u_2 & u_3 \\ \begin{bmatrix} k_b & -k_b \\ -k_b & k_b \end{bmatrix} & \begin{matrix} u_2 \\ u_3 \end{matrix} \end{matrix}. \tag{2.13}$$

In both (2.12) and (2.13), the displacements u_1, u_2, and u_3 are shown to indicate which components of the stiffness matrix correspond to which displacements, and also to assist the build-up of the system stiffness matrix, as shown in (2.14).

$$[K] = \Sigma[k] = \begin{matrix} u_1 & u_2 & u_3 \\ \begin{bmatrix} k_a & -k_a & 0 \\ -k_a & k_a + k_b & -k_b \\ 0 & -k_b & k_b \end{bmatrix} & \begin{matrix} u_1 \\ u_2 \\ u_3 \end{matrix}. \end{matrix} \tag{2.14}$$

Equation (2.14) is obtained by constructing a 3×3 matrix, the three elements of the matrix corresponding to the three nodal displacements, and then by adding the stiffness components from (2.12) and (2.13) into the appropriate boxes.

Thus, the matrix of (2.14) may be assumed to be a large rectangular box with 9 'pigeon holes', each corresponding to a combination of nodal displacements, and these 'pigeon holes' are filled with stiffness components from the elemental stiffness matrices of (2.12) and (2.13). The components of stiffness matrices are known as *stiffness influence coefficients*.

From (2.14), it can be seen that the size of $[K]$ depends on the total number of nodal displacements of the entire structure, whereas the sizes of elemental stiffness matrices depends on the number of nodal displacements per element. In general, it can be concluded that the size of $[K]$ can become very large.

2.4 METHOD OF SOLUTION

Equation (2.11) cannot be solved as it stands, because $[K]$ ir singular and 'floating in space'. Hence, to solve (2.11) it will be necessary to constrain the structure.

Let us say, in this case, the spring of Fig. 2.3 is constrained at node 3, so that $u_3 = 0$. Substituting $u_3 = 0$ into (2.10) and partitioning, the following is obtained:

$$\begin{Bmatrix} Q_1 \\ Q_2 \\ \hline R \end{Bmatrix} = \begin{bmatrix} k_a & -k_a & \vdots & 0 \\ -k_a & k_a + k_b & \vdots & -k_b \\ \hline 0 & -k_b & \vdots & k_b \end{bmatrix} \begin{Bmatrix} u_1 \\ u_2 \\ \hline 0 \end{Bmatrix}$$

or,

$$\begin{Bmatrix} q_F \\ R \end{Bmatrix} = \begin{bmatrix} K_{11} & K_{12} \\ K_{21} & K_{22} \end{bmatrix} \begin{Bmatrix} u_F \\ 0 \end{Bmatrix} \qquad (2.15)$$

where

$\{q_F\}$ = a vector of loads corresponding to the free displacement positions

$\{u_F\}$ = a vector of free nodal displacements

$\{R\}$ = a vector of reactions corresponding to the 'fixed' displacement positions.

$$[K_{11}] = \begin{bmatrix} k_a & -k_a \\ -k_a & (k_a + k_b) \end{bmatrix} = \text{a stiffness matrix corresponding to the free displacements.}$$

$$[K_{12}] = \begin{bmatrix} 0 \\ -k_b \end{bmatrix} = [K_{21}]^T$$

$$[K_{22}] = [k_b].$$

Expanding (2.15), the following is obtained:

$$\{q_F\} = [K_{11}]\{u_F\}$$

or

$$\{u_F\} = [K_{11}]^{-1}\{q_F\} \qquad (2.16)$$

and

$$\{R\} = [K_{21}]\{u_F\} = [K_{21}][K_{11}]^{-1}\{q_F\}. \qquad (2.17)$$

Once $\{u_F\}$ are obtained from (2.16), the internal forces can be found from expressions such as those given by (2.5), and if required, the reactions $\{R\}$ can be calculated from (2.17).

For large problems, it is not usual to invert $[K_{11}]$, but simply to solve the resulting simultaneous equations, as described in Chapter 1.

For some cases, there may be some movement or subsidence at the constraints, so that $\{u_3\} = \{u_c\}$, where

$\{u_c\}$ = a vector of 'fixed' displacements at the constraints, due to subsidence or movement.

For such cases, (2.16) becomes

$$\{u_F\} = [K_{11}]^{-1}(\{q_F\} - [K_{12}]\{u_c\},$$

and (2.17) becomes

$$\{R\} = [K_{21}]\{u_F\} + [K_{22}]\{u_c\}.$$

It should be noted that both the elemental stiffness matrix (2.7) and the structural stiffness matrix (2.14) are symmetrical, and this can be explained by the *Clerk-Maxwell reciprocal theorem* [25], which states that, for a linear elastic structure, the deflection at node 1 due to a unit force at node 2 is equal to the deflection at node 2 due to a unit force at node 1.

Thus, all stiffness and flexibility matrices in structural analysis are symmetrical, and this property can be used to save considerable amounts of space in the computer store for large problems.

It should also be noted that the elements of the stiffness matrices are known as *stiffness influence coefficients*, and that the size and direction of influence coefficients are very important as they govern stiffness of a structure, in much the same manner as values of k for an individual spring.

For simple elements such as rods and beams, there is no particular difficulty in deriving stiffness matrices, but for more sophicticated elements, this is one of the major problems.

In recent times, use of the various energy theorems has been found extremely useful for this process, and some of these are described below. Application of some of these theorems to the finite element approach is done in Chapter 5.

2.5 THE PRINCIPLE OF VIRTUAL WORK

This principle states that if an elastic body under a system of external forces is given a small virtual displacement, then the increase of work done by the forces external to the body is equal to the increase in strain energy stored. If

P = applied load

$\bar{u}$ = small virtual displacement

and

$\bar{\varepsilon}$ = small virtual strain,

then from Fig. 2.4, the increase of internal strain energy

$\delta(U_e)$ = area of vertical trapezium = $\bar{\varepsilon} \times \sigma \times$ vol.

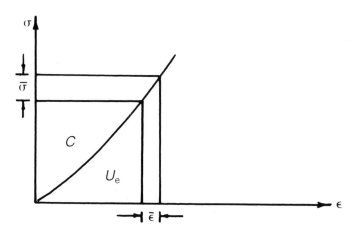

Fig. 2.4 – Stress-strain relationships.

Similarly, the external virtual work done (VW) is equal to the area given by the vertical trapezium of Fig. 2.5, that is,

$\delta(VW) = P\bar{u}$

but

$\delta(VW) = \delta(U_e)$

therefore

$P\bar{u} = \bar{\varepsilon}\sigma \times$ volume of body.

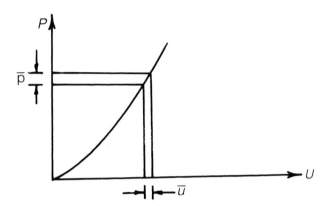

Fig. 2.5 – Force-displacement relationship.

In matrix form, the expressions for $\delta(VW)$ and $\delta(U_e)$ will appear, as in (2.18):

$$\{\bar{u}_i\}^{\mathrm{T}}\{P_i\} = \int\limits_{\mathrm{vol}} \{\bar{\varepsilon}\}^{\mathrm{T}}\{\sigma\}\,\mathrm{d}(\mathrm{vol}). \tag{2.18}$$

It is necessary for the multiplication of the matrices of (2.18) to be written as they are, as both work done and strain energy are scalars, that is,

$$\{\ \ \}^{\mathrm{T}}\{\ \ \} = \text{a scalar quantity.}$$

Now,

$$\{\sigma\} = [D]\{\varepsilon\}$$

where

$$[D] = \text{a matrix of elastic constants (see Chapter 5),}$$

therefore

$$\{\bar{u}_i\}^{\mathrm{T}}\{p\} = \int\limits_{\mathrm{vol}} \{\bar{\varepsilon}\}^{\bar{\mathrm{T}}}[D]\{\varepsilon\}\,\mathrm{d}(\mathrm{vol}).$$

For convenience, it is usual to obtain $\{\bar{\varepsilon}\}$ in terms of $\{\bar{u}_i\}$, and $\{\varepsilon\}$ in terms of $\{u_i\}$, so that

$$\int\limits_{\mathrm{vol}} \{\bar{\varepsilon}\}^{\mathrm{T}}[D]\{\varepsilon\}\,\mathrm{d}(\mathrm{vol}) = \{\bar{u}_i\}^{\mathrm{T}}[k]\{u_i\}$$

(see Chapter 5). Hence

$$\{\bar{u}_i\}^{\mathrm{T}}\{p_i\} = \{\bar{u}_i\}^{\mathrm{T}}[k]\{u_i\} \text{ or } \{p_i\} = [k]\{u_i\},$$

which is of similar form to (2.6).

Thus, this method is most suitable for stiffness analysis, and further work will be done in Chapter 5.

2.6   THE PRINCIPLE OF COMPLEMENTARY VIRTUAL WORK

Complementary strain energy and complementary virtual work are given by the horizontal trapeziums of Figs 2.4 and 2.5.

The principle of complementary virtual work states that if a body is subjected to a small virtual force, then the increase of complementary work done by the virtual force external to the body is equal to the increase of complementary strain energy.

From Fig. 2.5, complementary virtual work = area of horizontal trapezium = $u\bar{P}$. Similarly, complementary strain energy = $\varepsilon\bar{\sigma} \times$ vol. Equating these two, gives the following relationship:

$$u\bar{P} = \varepsilon\bar{\sigma} \times \text{vol},$$

which in matrix form appears as

$$\{\bar{p}_i\}^{\mathrm{T}} \{u_i\} = \int_{\text{vol}} \{\bar{\sigma}\}^{\mathrm{T}} \{\varepsilon\} \, \mathrm{d}(\text{vol})$$

where $\{\bar{\sigma}\}$ = a vector of virtual stresses, and $\{\bar{p}_i\}$ = a vector of virtual forces acting external to the element.

For this case, it is convenient to obtain $\{\bar{\varepsilon}\}$ and $\{\varepsilon\}$ in terms of $\{\bar{p}_i\}$ and $\{p_i\}$, respectively, so that

$$\int_{\text{vol}} \{\bar{\sigma}\}^{\mathrm{T}}\{\varepsilon\}d(vol) \sim \int_{\text{vol}} \{\bar{\varepsilon}\}^{\mathrm{T}} [D] \{\varepsilon\} d(vol) = \{\bar{u}_i\}^{\mathrm{T}} [k]\{u_i\}$$

but

$$\{u_i\} = [f]\{p_i\},$$

therefore

$$\int_{\text{vol}} \{\bar{\sigma}\}^{\mathrm{T}} \{\varepsilon\} \, \mathrm{d}(\text{vol}) = \{\bar{p}_i\}^{\mathrm{T}} [f] \{p_i\}.$$

Hence

$$\{\bar{p}_i\}^{\mathrm{T}}\{u_i\} = \{\bar{p}_i\}^{\mathrm{T}}[f] \{p_i\} \text{ or } \{u_i\} = [f]\{p_i\}$$

where

$$[f] = \text{flexibility matrix.}$$

From this, it can be seen that this method is suitable for flexibility analysis, but with some rearrangement of the equations, it can also be used for stiffness analysis.

2.7   METHOD OF MINIMUM POTENTIAL

This states that, to satisfy the equations of elasticity and equilibrium, the change of potential with respect to the displacement must be stationary. Let

π_p = total potential energy

U_e = strain energy

W = potential energy of load system

so that $\pi_p = U_e + W$. From Fig. 2.6, the strain energy stored in the body

$$U_e = \int_{vol} \text{area of vertical 'trapezium' } d(vol)$$

$$= \int_{vol} \int_0^\sigma \sigma \delta \varepsilon \, d(vol)$$

where

$\delta \varepsilon$ = increment of strain.

For linear elastic theory,

$$\varepsilon = \frac{\sigma}{E}$$

$$U_e = \int_{vol} \int_0^\sigma \frac{\sigma}{E} \, d\sigma \, d(vol)$$

$$= \int_{vol} \frac{\sigma^2}{2E} \, d(vol)$$

$$= \int_{vol} \frac{\sigma \varepsilon}{2} \, d(vol).$$

If the displacements of the body under concentrated loads P_i are u_i, then the potential energy of the load system is

$$W = -\Sigma P_i u_i.$$

Hence the total potential energy

$$\pi_p = \int_{vol} \frac{\sigma \varepsilon}{2} \, d(vol) - \Sigma P_i u_i. \qquad (2.19)$$

In matrix form these expressions appear as

$$U_e = \frac{1}{2} \int_{vol} \{\sigma\}^T \{\varepsilon\} d(vol)$$

$$= \frac{1}{2} \int_{vol} \{\varepsilon\}^T [D] \{\varepsilon\} d(vol)$$

$$= \frac{1}{2} \{u_i\}^T [k] \{u_i\}$$

and

$$W = -\{u_i\}^T \{p_i\}.$$

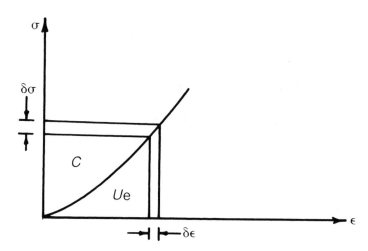

Fig. 2.6 – Stress-strain relationship.

Hence

$$\pi_p = \frac{1}{2} \{u_i\}^T [k] \{u_i\} - \{u_i\}^T \{p_i\}.$$

Now,

$$\frac{d\pi_p}{d \{u_i\}} = 0,$$

therefore

$$\{p_i\} = [k]\{u_i\},$$

that is, the method is suitable for stiffness analysis, and is discussed further in Chapter 5.

2.8   COMPLEMENTARY ENERGY THEOREM

From Fig. 2.7, the increase of complementary work done is given by the horizontal trapesium, that is,

$$\delta(C) = u\delta P$$

or

$$\frac{dC}{dP} = u.$$

Now,

$$U_e + C = Pu,$$

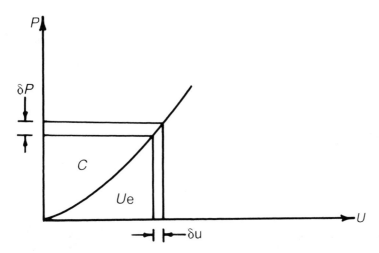

Fig. 2.7 – Load-displacement relationship.

therefore

$$C = Pu - U_e.$$

For linear elastic theory,

$$U_e = \int_{vol} \frac{\sigma\varepsilon}{2} \, d(vol).$$

In matrix form,

$$U_e = \frac{1}{2} \{p_i\}^T [f]\{p_i\},$$

therefore

$$C = \{p_i\}^T \{u_i\} - \frac{1}{2} \{P_i\}^T [f]\{p_i\}.$$

Now for minimum complementary energy,

$$\frac{dC}{d\{p_i\}} = \{u_i\} - [f]\{p_i\} = 0,$$

therefore

$$\{u_i\} = [f]\{p_i\},$$

that is, the method is suitable for flexibility analysis.

For a more thorough study of energy theorems, the reader is referred to reference [26].

Chapter 3

Static analysis of pin-jointed trusses

Solution of problems in this book are based on the matrix displacement method, which first appeared in the 1940s [1 to 3]. During that period, the matrix displacement method was more of academic interest than of practical application, as it depended on digital computers, which were scarce, unreliable, and very expensive. In any case, many structural analysts preferred an alternative matrix method, known as the *force method*.

The force method is sometimes called the *flexibility method*, as it depends on a flexibility matrix, whilst the displacement method is often called the *stiffness method*, as it depends on a stiffness matrix.

In general, at that time, the force method was preferred to the displacement method, because its requirements for digital computers were less, and also because its principles could be more readily explained in analytical terms. Nevertheless, in the aircraft industry, where the force method first gained popularity, teams of operators of mechanical calculators were required to implement it.

During the fifties and sixties, much progress was made with digital computer technology, and as a result of this, the popularity of the displacement method grew very rapidly, as it lent itself more readily for automatic computation than did the force method.

Today, the vast majority of matrix analysis of structures is carried out via the displacement method, and many sophisticated finite element packages are available for both large and small computers. The type of complex structure which defied theoretical analysis a few decades ago, can be solved on a microcomputer by a relative non-expert in structures.

In the present chapter, only pin-jointed trusses, loaded at their joints, as shown in Figs 3.1 and 3.2, will be considered. The elements of these trusses are called rods, and they possess only axial stiffness, rather similar to the springs of Chapter 2. The structures can be statically determinate or statically indeterminate, and they can be composed of members with different cross-sections and elastic moduli.

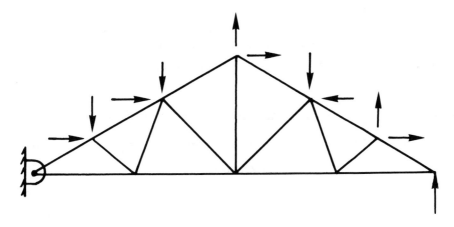

Fig. 3.1 – Plane pin-jointed truss.

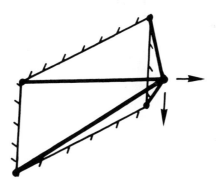

Fig. 3.2 – Pin-jointed space truss.

3.1 TO OBTAIN [k] FOR A ROD ELEMENT

If a uniform rod of length 'l', cross-sectional area 'A', and Young's modulus of elasticity 'E', is subjected to axial nodal forces 'X_1' and 'X_2', as shown in Fig. 3.3, the force-displacement relationships are:

$$X_1 = \frac{AE}{l}(u_1 - u_2) \tag{3.1}$$

$$X_2 = -X_1$$

$$= \frac{AE}{l}(u_2 - u_1) \tag{3.2}$$

where

$\quad u_1 =$ displacement at the nodal point (1) in the x direction

$\quad u_2 =$ displacement at the nodal point (2) in the x direction

$\quad X_1 =$ force at the nodal point (1) in the x direction

$\quad X_2 =$ force at the nodal point (2) in the x direction.

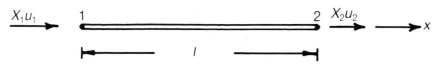

Fig. 3.3 – Rod element.

In matrix form equations (3.1) and (3.2) can be written

$$\left\{ \begin{matrix} X_1 \\ X_2 \end{matrix} \right\} = \frac{AE}{l} \begin{bmatrix} 1 & -1 \\ -1 & 1 \end{bmatrix} \left\{ \begin{matrix} u_1 \\ u_2 \end{matrix} \right\}$$

or

$$\{p_i\} = [k]\{u_i\} \tag{3.3}$$

where

$$[k] = \frac{AE}{l} \begin{bmatrix} 1 & -1 \\ -1 & 1 \end{bmatrix} \tag{3.4}$$

$\qquad =$ elemental stiffness matrix for a rod in local coordinates

$$\{p_i\} = \left\{ \begin{matrix} X_1 \\ X_2 \end{matrix} \right\} = \text{a vector representing forces acting on the element at its nodes}$$

$$\{u_i\} = \left\{ \begin{matrix} u_1 \\ u_2 \end{matrix} \right\} = \text{a vector representing displacements of the element at its nodes.}$$

3.2 TO OBTAIN $[k^o]$ FOR A TWO-DIMENSIONAL ROD ELEMENT

The elemental stiffness matrix (3.4) is, in general, not very useful in its one-dimensional form, as most members of a truss will be inclined at various angles to the horizontal. Hence, it is necessary to obtain a general two-dimensional form of $[k]$, which can cater for a rod inclined at any angle. Let

$[k^o]$ = elemental stiffness matrix for a two-dimensional rod in global
   coordinates.

The global axes Ox^o and Oy^o are shown in Fig. 3.4, together with the local axes of the rod-Ox and Oy. From Fig. 3.4, it can be seen that the relationship between the local displacements u and v, and the global displacements u^o and v^o, are given by (3.5):

$$v = v^o \cos \alpha - u^o \sin \alpha$$
$$u = u^o \cos \alpha + v^o \sin \alpha \qquad (3.5)$$

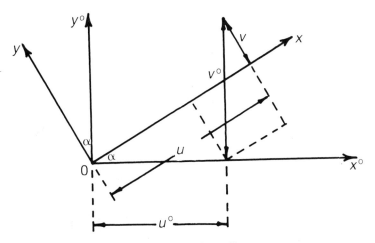

Fig. 3.4 – Global and local coordinate systems.

In matrix form equation (3.5) can be written

$$\begin{Bmatrix} u \\ v \end{Bmatrix} = \begin{bmatrix} \cos \alpha & \sin \alpha \\ -\sin \alpha & \cos \alpha \end{bmatrix} \begin{bmatrix} u^o \\ v^o \end{bmatrix}$$

or

$$\{u_i\} = [\zeta] \{u^o_i\} \qquad (3.6)$$

As the rod has two nodal points the relationship between global and local displacement is

$$
\begin{Bmatrix} u_1 \\ v_1 \\ u_2 \\ v_2 \end{Bmatrix} = \begin{bmatrix} \zeta & 0 \\ \hline 0 & \zeta \end{bmatrix} \begin{Bmatrix} u_1^{\,o} \\ v_1^{\,o} \\ u_2^{\,o} \\ v_2^{\,o} \end{Bmatrix}
\tag{3.7}
$$

or

$$
\{u_i\} = \begin{bmatrix} | - | \\ \hline | - | \end{bmatrix} \{u_i^o\}.
\tag{3.8}
$$

From section 1.5.5,

$$
\begin{bmatrix} | - | \\ \hline | - | \end{bmatrix}
$$

can be seen to be orthogonal, hence

$$
\begin{bmatrix} | - | \\ \hline | - | \end{bmatrix}^{\mathrm{T}} = \begin{bmatrix} | - | \\ \hline | - | \end{bmatrix}^{-1}
\tag{3.9}
$$

By a process similar to the above, it can be shown that

$$
\{p_i\} = \begin{bmatrix} | - | \\ \hline | - | \end{bmatrix} \{p_i^o\}
\tag{3.10}
$$

where

$$
\{p_i\}^{\mathrm{T}} = [X_1\ Y_1\ X_2\ Y_2] \text{ and } \{p_i^o\}^{\mathrm{T}} = [X_1^{\,o}\ Y_1^{\,o}\ X_2^{\,o}\ Y_2^{\,o}].
$$

Substituting equations (3.8) and (3.10) into equation (3.3) gives

$$
\begin{bmatrix} | - | \\ \hline | - | \end{bmatrix} \{p_i^o\} = [k] \begin{bmatrix} | - | \\ \hline | - | \end{bmatrix} \{u_i^o\}
\tag{3.11}
$$

and using the property of equation (3.9)

$$
[p_i^o\} = \begin{bmatrix} | - | \\ \hline | - | \end{bmatrix}^{\mathrm{T}} [k] \begin{bmatrix} | - | \\ \hline | - | \end{bmatrix} \{u_i^o\}
\tag{3.12}
$$

$$= [k^\circ] \{u^\circ_i\} \tag{3.13}$$

or

$$[k^\circ] = \begin{bmatrix} | & —— & | \\ —— & —— & —— \\ | & —— & | \end{bmatrix}^{\mathsf{T}} [k] \begin{bmatrix} | & —— & | \\ —— & —— & —— \\ | & —— & | \end{bmatrix} \tag{3.14}$$

= elemental stiffness matrix in global coordinates.

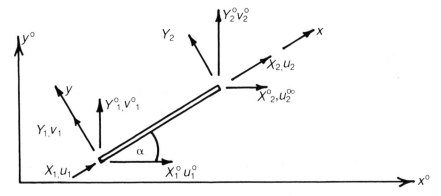

Fig. 3.5 – Rod element in local and global coordinate systems.

From Fig. 3.5, it can be seen that

$$\begin{Bmatrix} X_1 \\ Y_1 \\ X_2 \\ Y_2 \end{Bmatrix} = [k] \{u_i\} = \frac{AE}{l} \begin{bmatrix} 1 & 0 & -1 & 0 \\ 0 & 0 & 0 & 0 \\ -1 & 0 & 1 & 0 \\ 0 & 0 & 0 & 0 \end{bmatrix} \begin{Bmatrix} u_1 \\ v_1 \\ u_2 \\ v_2 \end{Bmatrix} \tag{3.15}$$

where, X_1 and X_2 are forces in the x direction, and are given by (3.1) and (3.2), and Y_1 and Y_2 are forces in the y direction and are equal to zero.

Substituting the relevant parts of equations (3.6), (3.7), and (3.15) into equation (3.14), the elemental stiffness matrix for a rod in Global coordinates is given by (3.16):

$$[k^\circ] = \frac{AE}{l} \begin{matrix} \begin{matrix} u_1^{\,\circ} & v_1^{\,\circ} & u_2^{\,\circ} & v_2^{\,\circ} \end{matrix} \\ \begin{bmatrix} c^2 & cs & -c^2 & -cs \\ cs & s^2 & -cs & -s^2 \\ -c^2 & -cs & c^2 & cs \\ -cs & -s^2 & cs & s^2 \end{bmatrix} \end{matrix} \begin{matrix} u_1^{\,\circ} \\ v_1^{\,\circ} \\ u_2^{\,\circ} \\ v_2^{\,\circ} \end{matrix} \tag{3.16}$$

where

$$c = \cos \alpha$$
$$s = \sin \alpha.$$

If a rod is governed by the nodes 'i' and 'j', and if the global coordinates of these nodes are (x_i°, y_i°) and (x_j°, y_j°), then c, s, and l can be determined from the following expressions:

$$\left.\begin{array}{l} l = \sqrt{(x_j^\circ - x_i^\circ)^2 + (y_j^\circ - y_i^\circ)^2} \\ c = (x_j^\circ - x_i^\circ)/l \\ s = (y_j^\circ - y_i^\circ)/l. \end{array}\right\} \tag{3.17}$$

3.3   PLANE PIN-JOINTED TRUSSES

To demonstrate the method of application, the elemental stiffness matrix of (3.16) will be applied to the following two examples.

3.3.1   Example 3.1

Find the forces in the members of the plane pin-jointed truss shown in Fig. 3.6. The truss is composed of three uniform members numbered (1) to (3), of cross-sectional areas A_0, A_0, and $\sqrt{2}\, A_0$ respectively, and all of constant E. It may be assumed that there is no initial-lack-of-fit.

Member (1) will be known as member 1–2 as it lies between nodal points 1 and 2 and members (2) and (3) will be known as members 1–3 and 1–4 respectively, for similar reasons.

Member 1–2

$$x_1^\circ = 8.66 \text{ m} \qquad x_2^\circ = 0 \text{ m}$$
$$y_1^\circ = -5 \text{ m} \qquad y_2^\circ = 0 \text{ m}.$$

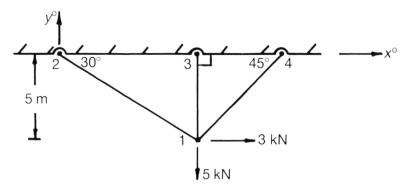

Fig. 3.6 – Plane pin-jointed truss.

From (3.17),

$$l_{1-2} = \sqrt{[(0 - 8.66)^2 + (0 + 5)^2]} = 10 \text{ m}$$

$$c = \cos \alpha = \frac{x_2{}^\circ - x_1{}^\circ}{l_{1-2}} = \frac{0 - 8.66}{10} = -\sqrt{3}/2$$

$$s = \sin \alpha = \frac{y_2{}^\circ - y_1{}^\circ}{l_{1-2}} = \frac{0 - (-5)}{10} = 1/2.$$

Substituting these values into (3.16), the elemental stiffness matrix for member 1–2, in global coordinates, is given by (3.18):

$$[k^\circ]_{1-2} = \frac{A_0 E}{10} \begin{bmatrix} \dfrac{3}{4} & -\dfrac{\sqrt{3}}{4} & -\dfrac{3}{4} & \dfrac{\sqrt{3}}{4} \\[2mm] -\dfrac{\sqrt{3}}{4} & \dfrac{1}{4} & \dfrac{\sqrt{3}}{4} & -\dfrac{1}{4} \\[2mm] -\dfrac{3}{4} & \dfrac{\sqrt{3}}{4} & \dfrac{3}{4} & -\dfrac{\sqrt{3}}{4} \\[2mm] \dfrac{\sqrt{3}}{4} & -\dfrac{1}{4} & -\dfrac{\sqrt{3}}{4} & \dfrac{1}{4} \end{bmatrix} \begin{matrix} u_1{}^\circ \\[2mm] v_1{}^\circ \\[2mm] u_2{}^\circ \\[2mm] v_2{}^\circ. \end{matrix} \qquad (3.18)$$

with columns labelled $u_1{}^\circ \quad v_1{}^\circ \quad u_2{}^\circ \quad v_2{}^\circ$

Member 1–3

$$x_3{}^\circ = 8.66 \text{ m} \qquad y_3{}^\circ = 0$$

From (3.17),

$$l_{1-3} = \sqrt{[(8.66 - 8.66)^2 + (0 + 5)^2]} = 5 \text{ m}$$

$$c = \frac{x_3{}^\circ - x_1{}^\circ}{l_{1-3}} = 0$$

$$s = \frac{y_3{}^\circ - y_1{}^\circ}{l_{1-3}} = 1.$$

Substituting these values of c, s, and l_{1-3} into (3.16), the elemental stiffness matrix for member 1–3, in global coordinates, is given by (3.19):

$$[k^\circ]_{1-3} = \frac{A_0 E}{10} \begin{bmatrix} 0 & 0 & 0 & 0 \\ 0 & 2 & 0 & -2 \\ 0 & 0 & 0 & 0 \\ 0 & -2 & 0 & 2 \end{bmatrix} \begin{matrix} u_1{}^\circ \\ v_1{}^\circ \\ u_3{}^\circ \\ v_3{}^\circ. \end{matrix} \qquad (3.19)$$

with columns labelled $u_1{}^\circ \quad v_1{}^\circ \quad u_3{}^\circ \quad v_3{}^\circ$

Member 1–4

$$x_4{}^\circ = 13.66 \text{ m} \qquad y_4{}^\circ = 0$$

From (3.17),

$$l_{1-4} = \sqrt{[(13.66 - 8.66)^2 + (0 + 5)^2]} = 5\sqrt{2} \text{ m}$$
$$c = s = 1/\sqrt{2}.$$

Substituting c, s, and l_{1-4} into (3.16), the elemental stiffness matrix for member 1–4, in global coordinates, is given by (3.20):

$$[k^\circ]_{1-4} = \frac{\sqrt{2}A_0 E}{5\sqrt{2}} \begin{bmatrix} \frac{1}{2} & \frac{1}{2} & -\frac{1}{2} & -\frac{1}{2} \\ \frac{1}{2} & \frac{1}{2} & -\frac{1}{2} & -\frac{1}{2} \\ -\frac{1}{2} & -\frac{1}{2} & \frac{1}{2} & \frac{1}{2} \\ -\frac{1}{2} & -\frac{1}{2} & \frac{1}{2} & \frac{1}{2} \end{bmatrix}. \qquad (3.20)$$

To obtain the structural stiffness matrix $[K^\circ]$ for the entire structure, the three elemental stiffness matrices (3.18) to (3.20) must be added into the appropriate parts of (3.21), as described in section 2, where it can be seen that the stiffness influence coefficients correspond to the appropriate displacements

	$u_1{}^\circ$	$v_1{}^\circ$	$u_2{}^\circ$	$v_2{}^\circ$
	1.750	0.567	−0.750	0.433
	0.567	3.250	0.433	−0.250
	−0.750	0.433	0.750	−0.433
	0.433	−0.250	−0.433	0.250
$[K^\circ] = \dfrac{A_0 E}{10}$	0.000	0.000	0.000	0.000
	0.000	−2.000	0.000	0.000
	−1.000	−1.000	0.000	0.000
	−1.000	−1.000	0.000	0.000

$u_3{}^\circ$	$v_3{}^\circ$	$u_4{}^\circ$	$v_4{}^\circ$	
0.000	0.000	−1.000	−1.000	$u_1{}^\circ$
0.000	−2.000	−1.000	−1.000	$v_1{}^\circ$
0.000	0.000	0.000	0.000	$u_2{}^\circ$
0.000	0.000	0.000	0.000	$v_2{}^\circ$
0.000	0.000	0.000	0.000	$u_3{}^\circ$
0.000	2.000	0.000	0.000	$v_3{}^\circ$
0.000	0.000	1.000	1.000	$u_4{}^\circ$
0.000	0.000	1.000	1.000	$v_4{}^\circ$. (3.21)

From (3.21) it can be seen that:

$$[K_{11}{}^{\circ}] = \frac{A_0 E}{10} \begin{matrix} u_1{}^{\circ} & v_1{}^{\circ} \\ \begin{bmatrix} 1.750 & 0.567 \\ 0.567 & 3.250 \end{bmatrix} & \begin{matrix} u_1{}^{\circ} \\ v_1{}^{\circ} \end{matrix} \end{matrix}$$

where

$[K_{11}]$ = structural stiffness matrix corresponding to the free displacements $u_1{}^{\circ}$ and $v_1{}^{\circ}$.

From (2.16),

$$\{u_F\} = [K_{11}{}^{\circ}]^{-1} \{q_F\}$$

where

$$\{u_F\} = \begin{Bmatrix} u_1{}^{\circ} \\ v_1{}^{\circ} \end{Bmatrix}$$

and

$$\{q_F\} = \begin{Bmatrix} 3 \\ -5 \end{Bmatrix}.$$

From (1.31),

$$[K_{11}{}^{\circ}]^{-1} = \frac{10}{A_0 E} \begin{bmatrix} 0.606 & -0.106 \\ -0.106 & 0.326 \end{bmatrix}$$

$$\begin{Bmatrix} u_1{}^{\circ} \\ v_1{}^{\circ} \end{Bmatrix} = \frac{10}{A_0 E} \begin{bmatrix} 0.606 & -0.106 \\ -0.106 & 0.326 \end{bmatrix} \begin{Bmatrix} 3 \\ -5 \end{Bmatrix}$$

$$= \frac{10}{A_0 E} \begin{Bmatrix} 2.348 \\ -1.948 \end{Bmatrix}.$$

The member forces F_{1-2}, F_{1-3}, and F_{1-4}, are determined as follows:

Member 1–2

$$c = -0.866 \qquad s = 0.5.$$

From (3.6),

$$\begin{Bmatrix} u_1 \\ v_1 \end{Bmatrix} = [\zeta] \begin{Bmatrix} u_1{}^{\circ} \\ v_1{}^{\circ} \end{Bmatrix}$$

$$= \begin{bmatrix} c & s \\ -s & c \end{bmatrix} \frac{10}{A_0 E} \begin{Bmatrix} 2.348 \\ -1.948 \end{Bmatrix}$$

$$= \begin{bmatrix} -0.866 & 0.5 \\ -0.5 & -0.866 \end{bmatrix} \frac{10}{A_0 E} \begin{Bmatrix} 2.348 \\ -1.948 \end{Bmatrix}$$

$$\begin{Bmatrix} u_1 \\ v_1 \end{Bmatrix} = \frac{10}{A_0 E} \begin{Bmatrix} -3 \\ 0.513 \end{Bmatrix}$$

where u_1 and v_1 are the local displacements for member 1–2, at node 2, as shown in Fig. 3.7.

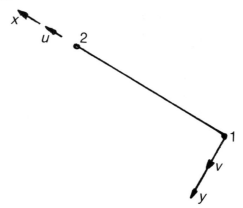

Fig. 3.7 – Member 1–2 in its local axes.

The displacements v_1 and v_2 are of no interest in this section, as they are perpendicular to member 1–2, and do not cause a change in the length of the member. The displacements u_1 and u_2 are in the axial direction of the member, and as they cause a change in the length of member 1–2, they are of much interest in determining the force in member 1–2.

From Hooke's law:

$$F_{1-2} = \frac{AE}{l}(u_2 - u_1) = \frac{A_0E}{10}\frac{10}{A_0E}(0 + 3)$$

$$F_{1-2} = 3 \text{ kN (tensile)}.$$

Member 1–3

$$c = 0 \qquad s = 1 \text{ (See Fig. (3.8))}.$$

From (3.6),

$$u_1 = \lfloor c \quad s \rfloor \begin{Bmatrix} u_1{}^o \\ v_1{}^o \end{Bmatrix}$$

$$u_1 = \lfloor 0 \quad 1 \rfloor \frac{10}{A_0E} \begin{Bmatrix} 2.348 \\ -1.948 \end{Bmatrix} = \frac{10}{A_0E}(-1.948).$$

From Hooke's law,

$$F_{1-3} = \frac{AE}{l}(u_3 - u_1)$$

$$= \frac{A_0E}{5}\left[0 - \frac{10}{A_0E}(-1.948)\right]$$

$$F_{1-3} = 3.896 \text{ kN (tensile)}$$

Fig. 3.8 – Member 1–3 in its local axes.

Member 1–4

$$c = 0.707 \qquad s = 0.707 \qquad \text{(See Fig. 3.9)}.$$

From (3.6),

$$u_1 = [0.707 \quad 0.707] \frac{10}{A_0 E} \left\{ \begin{matrix} 2.348 \\ -1.948 \end{matrix} \right\} = \frac{10}{A_0 E}(0.283).$$

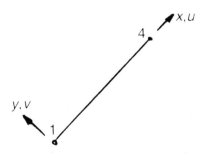

Fig. 3.9 – Member 1–4 in its local axes.

From Hooke's law,

$$F_{1-4} = \sqrt{2} \frac{A_0 E}{7.07} \left[0 - \frac{10}{A_0 E}(0.283) \right]$$

$$\underline{F_{1-4} = -0.566 \text{ kN (compressive)}}$$

N.B. It is of course perfectly acceptable to refer to a member by its high node first and its low node second, but if this is done, the elemental stiffness matrix may be different, although the structural stiffness matrix will be the same.

For example, if member 1–4 were referred to as member 4–1, the local origin of the member would be at 4 and the member would be pointing in the direction from 4 to 1, as shown in Fig. 3.10. The elemental stiffness matrix will be computed as follows:

Member 4–1

$$l_{4-1} = 7.07$$
$$c = -0.707 \qquad s = -0.707$$

so that,

$$
[k_{4-1}{}^\circ] = \begin{array}{cccc} u_4{}^\circ & v_4{}^\circ & u_1{}^\circ & v_1{}^\circ \\ \left[\begin{array}{cccc} 1 & 1 & -1 & -1 \\ 1 & 1 & -1 & -1 \\ -1 & -1 & 1 & 1 \\ -1 & -1 & 1 & 1 \end{array}\right] & & & \begin{array}{c} u_4{}^\circ \\ v_4{}^\circ \\ u_1{}^\circ \\ v_1{}^\circ \end{array} \end{array}
$$

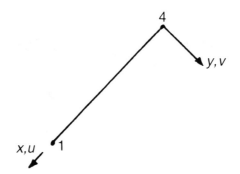

Fig. 3.10 – Member 4–1 in its local axes.

The structural stiffness matrix $[K_{11}{}^\circ]$ will not be changed, nor will the values of the computed nodal displacements.

To determine F_{4-1}

$$u_1 = \lfloor c \quad s \rfloor \left\{ \begin{array}{c} u_1{}^\circ \\ v_1{}^\circ \end{array} \right\} = [-0.707 \quad -0.707] \frac{10}{A_0 E} \left\{ \begin{array}{c} 2.348 \\ -1.948 \end{array} \right\}$$

$$= -0.283 \left(\frac{10}{A_0 E} \right)$$

$$F_{4-1} = \frac{\sqrt{2} A_0}{7.07} (u_1 - u_4) = -0.566 \text{ kN (as before).}$$

3.3.2 Example 3.2

Determine the forces in the members of the plane pin-jointed truss of Fig. 3.11. It may be assumed that for all members, $E = 2 \times 10^{11}$ N/m^2 and $A = 0.001$ m^2.

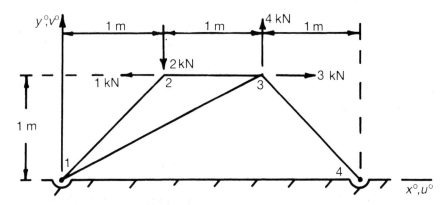

Fig. 3.11.

Member 1–2

$$x_1{}^\circ = 0 \qquad y_1{}^\circ = 0 \qquad x_2{}^\circ = 1 \qquad y_2{}^\circ = 1.$$

From (3.17),

$$l_{1-2} = \sqrt{(1-0)^2 + (1-0)^2} = 1.414$$
$$c = 0.707 \qquad s = 0.707.$$

Substitution of c, s, and l_{1-2} into (3.16) gives the following for the elemental stiffness matrix, in global coordinates, for member 1–2:

$$[k_{1-2}{}^\circ] =$$

$$
\begin{array}{cccc}
u_1{}^\circ & v_1{}^\circ & u_2{}^\circ & v_2{}^\circ \\
\end{array}
$$

$$
\begin{bmatrix}
70710.6781 & 70710.6781 & -70710.6781 & -70710.6781 \\
70710.6781 & 70710.6781 & -70710.6781 & -70710.6781 \\
-70710.6781 & -70710.6781 & 70710.6781 & 70710.6781 \\
-70710.6781 & -70710.6781 & 70710.6781 & 70710.6781
\end{bmatrix}
\begin{array}{l}
u_1{}^\circ \\
v_1{}^\circ \\
u_2{}^\circ \\
v_2{}^\circ.
\end{array}
$$

$$(3.22)$$

Member 2–3

$$x_3{}^\circ = 2 \qquad y_3{}^\circ = 1$$

$$l_{2-3} = \sqrt{(2-1)^2 + (1-1)^2} = 1; \quad c = 1; \quad s = 0.$$

Hence,

$$
[k_{2-3}{}^\circ] =
\begin{array}{c}
\begin{array}{cccc} u_2{}^\circ & v_2{}^\circ & u_3{}^\circ & v_3{}^\circ \end{array} \\
\left[
\begin{array}{cccc}
2E5 & 0 & -2E5 & 0 \\
0 & 0 & 0 & 0 \\
-2E5 & 0 & 2E5 & 0 \\
0 & 0 & 0 & 0
\end{array}
\right]
\begin{array}{c} u_2{}^\circ \\ v_2{}^\circ \\ u_3{}^\circ \\ v_3{}^\circ. \end{array}
\end{array}
\qquad (3.23)
$$

Member 1–3

$l_{1-3} = 2.236$ m; $c = 0.8945$; $s = 0.4472$.

Hence,

$[k_{1-3}{}^\circ] =$

$$
\begin{array}{c}
\begin{array}{cccc} u_1{}^0 & v_1{}^\circ & u_3{}^\circ & v_3{}^\circ \end{array} \\
\left[
\begin{array}{cccc}
71554.1753 & 35777.0876 & -71554.1753 & -35777.0876 \\
35777.0876 & 17888.5438 & -35777.0876 & -17888.5438 \\
-71554.1753 & -35777.0876 & 71554.1753 & 35777.0876 \\
-35777.0876 & -17888.5438 & 35777.0876 & 17888.5438
\end{array}
\right]
\begin{array}{c} u_1{}^\circ \\ v_1{}^\circ \\ u_3{}^\circ \\ v_3{}^\circ. \end{array}
\end{array}
$$

$$(3.24)$$

Member 3–4

$x_4{}^\circ = 3$; $y_4{}^\circ = 0$; $l_{3-4} = 1.414$; $c = 0.707$; $s = -0.707$.

Hence,

$[k_{3-4}{}^\circ] =$

$$
\begin{array}{c}
\begin{array}{cccc} u_3{}^4 & v_3{}^\circ & u_4{}^\circ & v_4{}^\circ \end{array} \\
\left[
\begin{array}{cccc}
70710.6781 & -70710.6781 & -70710.6781 & 70710.6781 \\
-70710.6781 & 70710.6781 & 70710.6781 & -70710.6781 \\
-70710.6781 & 70710.6781 & 70710.6781 & -70710.6781 \\
70710.6781 & -70710.6781 & -70710.6781 & 70710.6781
\end{array}
\right]
\begin{array}{c} u_3{}^\circ \\ v_3{}^\circ \\ u_4{}^\circ \\ v_4{}^\circ. \end{array}
\end{array}
$$

$$(3.25)$$

As the structure is firmly pinned at nodes 1 and 4, $u_1{}^\circ = v_1{}^\circ = u_4{}^\circ = v_4{}^\circ = 0$, so all the columns and rows relating to these can be ignored in determining $[K_{11}]$, the structural stiffness matrix corresponding to the free displacements. Hence, constructing a 4×4 matrix, as shown in (3.26), $[K_{11}]$ can be assembled from the four elemental stiffness matrices (3.22) to (3.25) by adding together the components of the stiffness coefficients with reference to the free displacements $u_2{}^\circ$, $v_2{}^\circ$, $u_3{}^\circ$, and $v_3{}^\circ$.

$$[K_{11}] = \begin{array}{cccc} u_2^{\circ} & v_2^{\circ} & u_3^{\circ} & v_3^{\circ} \\ \left[\begin{array}{c|c|c|c} 270711 & 70711 & -200000 & 0 \\ \hline 70711 & 70711 & 0 & 0 \\ \hline -200000 & 0 & 342265 & -34934 \\ \hline 0 & 0 & -34934 & 88599 \end{array} \right] & \begin{array}{c} u_2^{\circ} \\ \\ v_2^{\circ} \\ \\ u_3^{\circ} \\ \\ v_3^{\circ}. \end{array} \end{array} \qquad (3.26)$$

The vector of nodal loads, corresponding to the free displacements, is given by:

$$\{q_F\} \begin{Bmatrix} -1 \\ -2 \\ 3 \\ 4 \end{Bmatrix}.$$

Solving the four simultaneous equations, results in the following for the free displacements:

$$\{u_F\} = \begin{Bmatrix} u_2^{\circ} \\ v_2^{\circ} \\ u_3^{\circ} \\ v_3^{\circ} \end{Bmatrix} = \begin{Bmatrix} 4.8405\text{E-5} \\ -7.6689\text{E-5} \\ 4.3405\text{E-5} \\ 6.2261\text{E-5} \end{Bmatrix}.$$

By determining the local displacements u_2 and u_3 for each member and using Hooke's law, the forces in the members can be found to be as follows:

$F_{1-2} = -2.828$ kN (compressive)

$F_{2-3} = -1$ kN (compressive)

$F_{1-3} = 5.963$ kN (tensile)

$F_{3-4} = 1.886$ kN (tensile).

3.4   TO OBTAIN [k°] FOR A THREE-DIMENSIONAL ROD

In local coordinates, for an axially loaded rod in space, equations (3.1) and (3.2) become:

$$\begin{Bmatrix} X_1 \\ Y_1 \\ Z_1 \\ X_2 \\ Y_2 \\ Z_2 \end{Bmatrix} = [k]\{u_i\}$$

where

$$\{u_i\}^\mathrm{T} = [u_1 \quad v_1 \quad w_1 \quad u_2 \quad v_2 \quad w_2]$$

X_1 and X_2 are nodal forces in the x direction

Y_1 and Y_2 are nodal forces in the y direction and are equal to zero

Z_1 and Z_2 are nodal forces in the z direction and are equal to zero

$$[k] = \frac{AE}{l} \begin{bmatrix} 1 & 0 & 0 & -1 & 0 & 0 \\ 0 & 0 & 0 & 0 & 0 & 0 \\ 0 & 0 & 0 & 0 & 0 & 0 \\ -1 & 0 & 0 & 1 & 0 & 0 \\ 0 & 0 & 0 & 0 & 0 & 0 \\ 0 & 0 & 0 & 0 & 0 & 0 \end{bmatrix}. \tag{3.27}$$

In local coordinates, $[k]$ is not particularly useful, as in practice most rods will be inclined at some angle, as shown in Fig. 3.12.

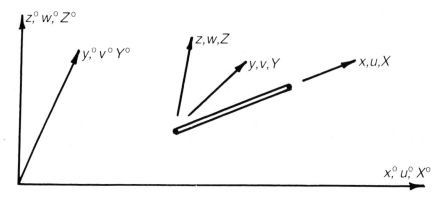

Fig. 3.12 – Rod element in three-dimensional coordinate systems.

By resolution, it can be shown that,

$$X = X^\circ Cx, x^\circ + Y^\circ Cx, y^\circ + Z^\circ Cx, z^\circ$$
$$Y = X^\circ Cy, x^\circ + Y^\circ Cy, y^\circ + Z^\circ Cy, z^\circ$$
$$Z = X^\circ Cz, x^\circ + Y^\circ Cz, y^\circ + Z^\circ Cz, z^\circ. \tag{3.28}$$

Similarly,

$$\begin{Bmatrix} u \\ v \\ w \end{Bmatrix} = \begin{bmatrix} Cx, x^\circ & Cx, y^\circ & Cx, z^\circ \\ Cy, x^\circ & Cy, y^\circ & Cy, z^\circ \\ Cz, x^\circ & Cz, y^\circ & Cz, z^\circ \end{bmatrix} \begin{Bmatrix} u^\circ \\ v^\circ \\ w^\circ \end{Bmatrix} \tag{3.29}$$

where

$x-y-z$ = local axes of rod element

x^o–y^o–z^o = global axes of system or structure

X = force in x direction

Y = force in y direction

Z = force in z direction

X^o = force in x^o direction

Y^o = force in y^o direction

Z^o = force in z^o direction

Cx, x^o = direction cosine of ox with ox^o

Cx, y^o = direction cosine of ox with oy^o

Cx, z^o = direction cosine of ox with oz^o

Cy, x^o, etc. = direction cosine of oy with ox^o, etc.

Applying equation (3.28) to the two nodes of the rod, the following matrix expression is obtained:

$$\{P_i\} = \left[\begin{array}{c|c} \rule{1cm}{0pt} \\ \hline - \\ \hline \rule{1cm}{0pt} \end{array}\right] \{P_i^o\} \qquad (3.30)$$

where

$$\{P_i\}^T = [X_1 \quad Y_1 \quad Z_1 \quad X_2 \quad Y_2 \quad Z_2]$$
$$\{P_i^o\}^T = [X_1^o \quad Y_1^o \quad Z_1^o \quad X_2^o \quad Y_2^o \quad Z_2^o]$$

$$\left[\begin{array}{c|c} \rule{1cm}{0pt} \\ \hline - \\ \hline \rule{1cm}{0pt} \end{array}\right] = \left[\begin{array}{c|c} \zeta & O_3 \\ \hline O & \zeta \end{array}\right] \qquad (3.31)$$

$$[\zeta] = \begin{bmatrix} Cx, x^o & Cx, y^o & Cx, z^o \\ Cy, x^o & Cy, y^o & Cy, z^o \\ Cz, x^o & Cz, y^o & Cz, z^o \end{bmatrix}.$$

Now,

$$[k^o] = \left[\begin{array}{c|c} \rule{1cm}{0pt} \\ \hline - \\ \hline \rule{1cm}{0pt} \end{array}\right]^T [k] \left[\begin{array}{c|c} \rule{1cm}{0pt} \\ \hline - \\ \hline \rule{1cm}{0pt} \end{array}\right].$$

Hence, by substitution of (3.27) and (3.31) into (3.14), the following is obtained for [k^o], the elemental stiffness matrix for a three-dimensional rod in global coordinates:

$$
[k^\circ] = \frac{AE}{l}
\begin{bmatrix}
\begin{array}{ccc}
u_1^\circ & v_1^\circ & w_1^\circ
\end{array} \\
\begin{array}{ccc}
C^2x, x^\circ & & \\
Cx, x^\circ, Cx, y^\circ & C^2x, y^\circ & \\
Cx, x^\circ Cx, z^\circ & Cx, y^\circ Cx, z^\circ & C^2x, z^\circ \\
\hline
-C^2x, x^\circ & -Cx, x^\circ Cx, y^\circ & -Cx, x^\circ Cx, z^\circ \\
-Cx, x^\circ Cx, y^\circ & -C^2x, y^\circ & -Cx, y^\circ Cx, z^\circ \\
-Cx, x^\circ Cx, z^\circ & -Cx, y^\circ Cx, z^\circ & -C^2x, z^\circ
\end{array}
\end{bmatrix}
$$

$$
\begin{array}{ccc}
u_2^\circ & v_2^\circ & w_2^\circ
\end{array}
$$

Symmetrical

$$
\begin{array}{ccc}
 & & \\
\hline
C^2x, x^\circ & & \\
Cx, x^\circ Cx, y^\circ & C^2x, y^\circ & \\
Cx, x^\circ Cx, z^\circ & Cx, y^\circ Cx, z^\circ C^2x, z^\circ
\end{array}
\quad
\begin{array}{c}
u_1^\circ \\ v_1^\circ \\ w_1^\circ \\ \\ u_2^\circ \\ v_2^\circ \\ w_2^\circ
\end{array}
\tag{3.32}
$$

which can be written in the form:

$$
[k^\circ] =
\left[
\begin{array}{c|c}
a & -a \\
\hline
-a & a
\end{array}
\right]
$$

where

$$
[a] = \frac{AE}{l}
\begin{bmatrix}
C^2x, x^\circ & & \text{Symmetrical} \\
Cx, x^\circ Cx, y^\circ & C^2x, y^\circ & \\
Cx, x^\circ Cx, z^\circ & Cx, y^\circ Cx, z^\circ & C^2x, z^\circ
\end{bmatrix}
$$

$$
\left.
\begin{aligned}
l &= \sqrt{[(x_j^\circ - x_i^\circ)^2 + (y_j^\circ - y_i^\circ)^2 + (z_j^\circ - z_i^\circ)^2]} \\
Cx, x^\circ &= (x_j^\circ - x_i^\circ)/l \\
Cx, y^\circ &= (y_j^\circ - y_i^\circ)/l \\
Cx, z^\circ &= (z_j^\circ - z_i^\circ)/l
\end{aligned}
\right\}
\tag{3.33}
$$

3.5   PIN-JOINTED SPACE TRUSSES

The solution for these follows a similar pattern to that for section 3.3, except that $[k^\circ]$ and $[\Xi]$ are used from (3.31) and (3.32). To demonstrate the method, two worked examples will be considered.

3.5.1 Example 3.3

The tripod shown in Fig. 3.13(a) and 3.13(b) consists of three identical
members which are pinned together at the nodal point 4 and to a rigid base at
the nodal points 1, 2, and 3. Calculate the displacements at the nodal point 4
and the member forces.

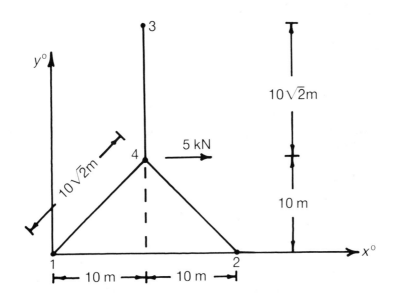

Fig. 3.13(a) – Plan of pin-jointed tripod.

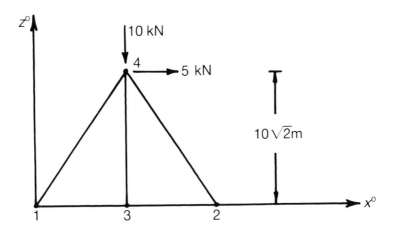

Fig. 3.13(b) – Front elevation of pin-jointed tripod, looking in y^o direction.

Member 1–4

$$x_1{}^\circ = y_1{}^\circ = z_1{}^\circ = 0; \quad x_4{}^\circ = y_4{}^\circ = 10 \text{ m}; \quad z_4{}^\circ = 10\sqrt{2} \text{ m}.$$

From (3.33),

$$l = \sqrt{(10^2 + 10^2 + 10^2 \times 2)} = 20 \text{ m}$$

$$Cx, x^\circ = \frac{x_4{}^\circ - x_1{}^\circ}{l} = \frac{10}{20} = \frac{1}{2}$$

$$Cx, y^\circ = \frac{y_4{}^\circ - y_1{}^\circ}{l} = \frac{1}{2}$$

$$Cx, z^\circ = \frac{z_4{}^\circ - z_1{}^\circ}{l} = \frac{10\sqrt{2}}{20} = \frac{1}{\sqrt{2}}.$$

Substituting the above values of l, Cx, x°, Cx, y° and Cx, z° into (3.32), the following is obtained for the elemental stiffness matrix for member 1–4 in global coordinates.

$$[k^\circ]_{1-4} = \frac{AE}{l}$$

$$
\begin{array}{cccccc}
u_1{}^\circ & v_1{}^\circ & w_1{}^\circ & u_4{}^\circ & v_4{}^\circ & w_4{}^\circ
\end{array}
$$

$$
\left[
\begin{array}{cccccc}
\dfrac{1}{4} & & & & & \\[2mm]
\dfrac{1}{4} & \dfrac{1}{4} & & & \text{Symmetrical} & \\[2mm]
\dfrac{1}{2\sqrt{2}} & \dfrac{1}{2\sqrt{2}} & \dfrac{1}{2} & & & \\[2mm]
-\dfrac{1}{4} & -\dfrac{1}{4} & -\dfrac{1}{2\sqrt{2}} & \dfrac{1}{4} & & \\[2mm]
-\dfrac{1}{4} & -\dfrac{1}{4} & -\dfrac{1}{2\sqrt{2}} & \dfrac{1}{4} & \dfrac{1}{4} & \\[2mm]
-\dfrac{1}{2\sqrt{2}} & -\dfrac{1}{2\sqrt{2}} & -\dfrac{1}{2} & \dfrac{1}{2\sqrt{2}} & \dfrac{1}{2\sqrt{2}} & \dfrac{1}{2}
\end{array}
\right]
\begin{array}{l}
u_1{}^\circ \\[2mm]
v_1{}^\circ \\[2mm]
w_1{}^\circ \\[2mm]
u_4{}^\circ \\[2mm]
v_4{}^\circ \\[2mm]
w_4{}^\circ.
\end{array}
\qquad (3.34)
$$

Member 2–4

$$x_2{}^\circ = 20 \text{ m}, \ y_2{}^\circ = 0, \ z_2{}^\circ = 0.$$

In a manner similar to that used for member 1–4, the following is obtained:
From (3.33),

$$l = 20 \text{ m}; \ Cx, x^\circ = -0.5; \ Cx, y^\circ = 0.5; Cx, z^\circ = 0.707.$$

Substituting the above values for l, Cx, x°, Cx, y° and Cx, z° into (3.32), the elemental stiffness matrix for member 2–4 in global coordinates is given by:

$$
[k^\circ]_{2-4} = \frac{AE}{l}
\begin{array}{cccccc}
u_2^\circ & v_2^\circ & w_2^\circ & u_4^\circ & v_4^\circ & w_4^\circ
\end{array}
$$

$$
[k^\circ]_{2-4} = \frac{AE}{l}
\left[
\begin{array}{ccc|ccc}
\frac{1}{4} & & & & & \\
-\frac{1}{4} & \frac{1}{4} & & & \text{Symmetrical} & \\
-\frac{1}{2\sqrt{2}} & \frac{1}{2\sqrt{2}} & \frac{1}{2} & & & \\
\hline
-\frac{1}{4} & \frac{1}{4} & \frac{1}{2\sqrt{2}} & \frac{1}{4} & & \\
\frac{1}{4} & -\frac{1}{4} & -\frac{1}{2\sqrt{2}} & -\frac{1}{4} & \frac{1}{4} & \\
\frac{1}{2\sqrt{2}} & -\frac{1}{2\sqrt{2}} & -\frac{1}{2} & -\frac{1}{2\sqrt{2}} & \frac{1}{2\sqrt{2}} & \frac{1}{2}
\end{array}
\right]
\begin{array}{c}
u_2^\circ \\
v_2^\circ \\
w_2^\circ \\
u_4^\circ \\
v_4^\circ \\
w_4^\circ.
\end{array}
$$

$$(3.35)$$

Member 3–4

$$x_3^\circ = 10 \text{ m}; \ y_3^\circ = 10 + 10\sqrt{2} \text{ m} = 24.14 \text{ m}; \ z_3^\circ = 0.$$

From (3.33),

$$l = 20 \text{ m}; \ Cx, x^\circ = 0; \ Cx, y^\circ = -0.707; \ Cx, z^\circ = 0.707.$$

Hence, by substituting the above values into (3.32),

$$
\begin{array}{cccccc}
u_3^\circ & v_3^\circ & w_3^\circ & u_4^\circ & v_4^\circ & w_4^\circ
\end{array}
$$

$$
[k^\circ]_{3-4} = \frac{AE}{l}
\left[
\begin{array}{ccc|ccc}
0 & & & & & \\
0 & \frac{1}{2} & & & \text{Symmetrical} & \\
0 & -\frac{1}{2} & \frac{1}{2} & & & \\
\hline
0 & 0 & 0 & 0 & & \\
0 & -\frac{1}{2} & \frac{1}{2} & 0 & \frac{1}{2} & \\
0 & \frac{1}{2} & -\frac{1}{2} & 0 & -\frac{1}{2} & \frac{1}{2}
\end{array}
\right]
\begin{array}{c}
u_3^\circ \\
v_3^\circ \\
w_3^\circ \\
u_4^\circ \\
v_4^\circ \\
w_4^\circ.
\end{array}
$$

$$(3.36)$$

Now, as $u_1{}^\circ = v_1{}^\circ = w_1{}^\circ = u_2{}^\circ = v_2{}^\circ = w_2{}^\circ = u_3{}^\circ = v_3{}^\circ = w_3{}^\circ = 0$, the structural stiffness matrix relating to the free displacements, $[K_{11}]$, can be obtained by adding together those parts of the stiffness influence coefficients from (3.34) to (3.36), corresponding to the free displacements $u_4{}^\circ$, $v_4{}^\circ$, and $w_4{}^\circ$, as shown in (3.37):

$$[K_{11}] = \frac{AE}{l} \begin{bmatrix} (\frac{1}{4}+\frac{1}{4}) & & \text{Symmetrical} \\ (\frac{1}{4}-\frac{1}{4}) & (\frac{1}{4}+\frac{1}{4}+\frac{1}{2}) & \\ \left(\frac{1}{2\sqrt{2}}-\frac{1}{2\sqrt{2}}\right) & \left(\frac{1}{2\sqrt{2}}+\frac{1}{2\sqrt{2}}-\frac{1}{2}\right) & (\frac{1}{2}+\frac{1}{2}+1) \end{bmatrix} \begin{matrix} u_4{}^\circ \\ v_4{}^\circ \\ w_4{}^\circ \end{matrix}$$

$$\begin{matrix} u_4{}^\circ & v_4{}^\circ & w_4{}^\circ \end{matrix}$$

(3.37)

the inverse of which is:

$$[K_{11}]^{-1} = \frac{l}{AE} \begin{bmatrix} 2.0 & 0.0 & 0.0 \\ 0.0 & 1.029 & -0.142 \\ 0.0 & -0.142 & 0.686 \end{bmatrix}.$$

(3.38)

Now,

$$\{q_F\} = \begin{Bmatrix} 5 \\ 0 \\ -10 \end{Bmatrix}$$

therefore

$$\{u_F\} = \begin{Bmatrix} u_4{}^\circ \\ v_4{}^\circ \\ w_4{}^\circ \end{Bmatrix} = \frac{l}{AE} \begin{bmatrix} 2.0 & 0.0 & 0.0 \\ 0.0 & 1.029 & -0.142 \\ 0.0 & -0.142 & 0.686 \end{bmatrix} \begin{Bmatrix} 5 \\ 0 \\ -10 \end{Bmatrix}$$

$$= \frac{l}{AE} \begin{Bmatrix} 10.0 \\ 1.42 \\ -6.86 \end{Bmatrix}.$$

The member forces are now obtained by considering the change of length of each of the members.

Member 1–4

$$Cx, x^\circ = 0.5; \ Cx, y^\circ = 0.5; \ Cx, z^\circ = 0.707,$$

From (3.29), therefore,

$$u_4 = \lfloor Cx, x^o \quad Cx, y^o \quad Cx, z^o \rfloor \begin{Bmatrix} u_4{}^o \\ v_4{}^o \\ w_4{}^o \end{Bmatrix}$$

$$= \lfloor 0.5 \quad 0.5 \quad 0.707 \rfloor \frac{l}{AE} \begin{Bmatrix} 10 \\ 1.42 \\ -6.86 \end{Bmatrix} = 0.86 \, l/AE.$$

From Hooke's law,

$$F_{1-4} = \frac{AE}{l}(u_4 - u_1)$$

$$F_{1-4} = \frac{AE}{l} \times \frac{0.86l}{AE}$$

$$\underline{F_{1-4} = 0.86 \text{ kN}}$$

Member 2–4

$$Cx, x^o = -0.5; \ Cx, y^o = 0.5; \ Cx, z^o = 0.707.$$

From (3.29),

$$u_4 = \lfloor -0.5 \quad 0.5 \quad 0.707 \rfloor \frac{l}{AE} \begin{Bmatrix} 10 \\ 1.42 \\ -6.86 \end{Bmatrix} = -9.14 \, l/AE.$$

From Hooke's law,

$$F_{2-4} = \frac{AE}{l}(u_4 - u_2)$$

$$\underline{F_{2-4} = 9.14 \text{ kN}}$$

Member 3–4

$$Cx, x^o = 0; \ Cx, y^o = -0.707; \ Cx, z^o = 0.707.$$

From (3.29),

$$u_4 = -5.854 \, \frac{l}{AE}$$

From Hooke's law,

$$\underline{F_{3-4} = -5.854 \text{ kN}}$$

N.B. The local 'v' displacements are not of any interest in calculating member forces, as they do not change the length of a member.

3.5.2   Example 3.4

Determine the forces in the members of the pin-jointed space truss of
Fig. 3.14(a) and 3.14(b). All members have a cross-sectional area of 0.001 m²
and elastic modulus of 2×10^8 kN/m².

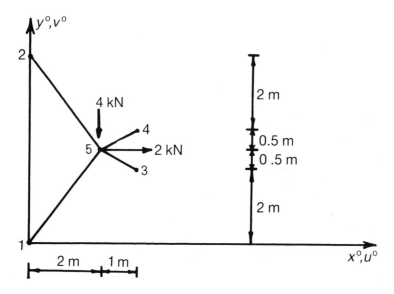

Fig. 3.14(a) – Plan of pin-jointed tripod.

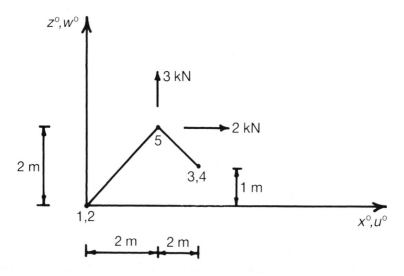

Fig. 3.14(b) – Front elevation of pin-jointed tripod.

Member 1–5

$$x_1{}^\circ = y_1{}^\circ = z_1{}^\circ = 0; \; x_5{}^\circ = 2 \text{ m } \; y_5{}^\circ = 2.5 \text{ m}; \; z_5{}^\circ = 2 \text{ m}$$

$$l = \sqrt{(2^2 + 2.5^2 + 2^2)} = 3.775 \text{ m}$$

$$Cx, x^\circ = \frac{x_5{}^\circ - x_1{}^\circ}{l} = 0.53$$

$$Cx, y^\circ = \frac{2.5}{3.775} = 0.662$$

$$Cx, z^\circ = 0.53$$

As $u_1{}^\circ = v_1{}^\circ = w_1{}^\circ = 0$, it is only necessary to calculate that part of the elemental stiffness matrix corresponding to the free displacements $u_5{}^\circ, v_5{}^\circ$, and $w_5{}^\circ$, as in (3.39).

$$[k_{1-5}] = \begin{array}{c} \\ \\ \\ \\ \end{array} \overset{\begin{array}{ccc} u_5{}^\circ & v_5{}^\circ & w_5{}^\circ \end{array}}{\begin{bmatrix} 14871.9 & 18589.9 & 14871.9 \\ 18589.9 & 23237.4 & 18589.9 \\ 14871.9 & 18589.9 & 14871.9 \end{bmatrix}} \begin{array}{c} u_5{}^\circ \\ v_5{}^\circ \\ w_5{}^\circ \end{array} \tag{3.39}$$

Member 2–5

$$x_2{}^\circ = 0; \; y_2{}^\circ = 5 \text{ m}; \; z_2{}^\circ = 0$$

$$l = 3.775 \text{ m}; \; Cx, x^\circ = 0.53; \; Cx, y^\circ = -0.662; \; Cx, z^\circ = 0.53.$$

As $u_2{}^\circ = v_2{}^\circ = w_2{}^\circ = 0$, it is only necessary to calculate that part of the elemental stiffness matrix corresponding to the free displacements $u_5{}^\circ, v_5{}^\circ$, and $w_5{}^\circ$, as in (3.40).

$$[k_{2-5}{}^\circ] = \overset{\begin{array}{ccc} u_5{}^\circ & v_5{}^\circ & w_5{}^\circ \end{array}}{\begin{bmatrix} 14871.9 & -18589.9 & 14871.9 \\ -18589.9 & 23237.4 & -18589.9 \\ 14871.9 & -18589.9 & 14871.9 \end{bmatrix}} \begin{array}{c} u_5{}^\circ \\ v_5{}^\circ \\ w_5{}^\circ \end{array} \tag{3.40}$$

Member 3–5

$$x_3{}^\circ = 3; \; y_3{}^\circ = 2; \; z_3{}^\circ = 1$$

$$l = \sqrt{((1-1)^2 + (0.5)^2 + 1^2)} = 1.5 \text{ m}$$

$$Cx, x^\circ = -1/1.5 = -0.6667; \; Cx, y^\circ = 0.3333; \; Cx, z^\circ = 0.6667.$$

Hence, by substitution,

$$[k_{3-5}] = \overset{\begin{array}{ccc} u_5{}^\circ & v_5{}^\circ & w_5{}^\circ \end{array}}{\begin{bmatrix} 59259.3 & -29629.6 & -59259.3 \\ -29629.0 & 14814.8 & 29629.6 \\ -59259.3 & 29629.6 & 59259.3 \end{bmatrix}} \begin{array}{c} u_5{}^\circ \\ v_5{}^\circ \\ w_5{}^\circ \end{array} \tag{3.41}$$

Member 4–5

$$x_4^{\,\circ} = 3;\ y_4^{\,\circ} = 3;\ z_4^{\,\circ} = 1$$
$$l = 1;\ Cx, x^{\circ} = -0.6667;\ Cx, y^{\circ} = -0.3333;\ Cx, z^{\circ} = 0.6667.$$

Hence, by substitution,

$$[k_{4-5}^{\,\circ}]\quad
\begin{array}{ccc}
u_5^{\,\circ} & v_5^{\,\circ} & w_5^{\,\circ}
\end{array}
$$
$$
\begin{bmatrix}
59259.3 & 29629.6 & -59259.3 \\
29629.6 & 14814.8 & -29629.6 \\
-59259.3 & -29629.6 & 59259.3
\end{bmatrix}
\begin{array}{l}
u_5^{\,\circ} \\
v_5^{\,\circ} \\
w_5^{\,\circ}.
\end{array}
\tag{3.42}
$$

By adding together the stiffness influence coefficients of the elemental stiffness matrices of (3.39) to (3.42), the structural stiffness matrix corresponding to the free displacements is obtained in (3.43).

$$[K_{11}^{\,\circ}]\quad
\begin{array}{ccc}
u_5^{\,\circ} & v_5^{\,\circ} & w_5^{\,\circ}
\end{array}
$$
$$
\begin{bmatrix}
148262 & 0 & -88775 \\
0 & 76104 & 0 \\
-88775 & 0 & 148262
\end{bmatrix}
\begin{array}{l}
u_5^{\,\circ} \\
v_5^{\,\circ} \\
w_5^{\,\circ}
\end{array}
\tag{3.43}
$$

and

$$\{q_F\} = \left\{ \begin{array}{c} 2 \\ -4 \\ 3 \end{array} \right\} \text{ kN.}$$

Solving the three simultaneous equations, the vector of free displacements is

$$\{u_F\} = \left\{ \begin{array}{c} u_5^{\,\circ} \\ v_5^{\,\circ} \\ w_5^{\,\circ} \end{array} \right\} = \left\{ \begin{array}{c} 3.992\text{E-}5 \\ -5.256\text{E-}5 \\ 4.413\text{E-}5 \end{array} \right\} \text{ m.}$$

The member forces are calculated as follows.

Member 1–5
From (3.29),

$$u_5 = \begin{bmatrix} 0.53 & 0.662 & 0.53 \end{bmatrix} \left\{ \begin{array}{c} 3.992\text{E-}5 \\ -5.256\text{E-}5 \\ 4.413\text{E-}5 \end{array} \right\} = 9.752\text{E-}6 \text{ m.}$$

From Hooke's law,

$$F_{1-5} = \frac{0.001 \times 2 \times 10^8}{3.775} \times 9.752\text{E-}6 = 0.515 \text{ kN.}$$

Member 2–5
From (3.29),

$$u_5 = 7.93\text{E-5 m}$$

therefore

$$F_{2\_5} = \frac{0.001 \times 2 \times 10^8 \times 7.93\text{E-5}}{3.775} = \underline{4.2 \text{ kN.}}$$

Member 3–5
From (3.29), $u_5 = -1.471\text{E-5}$
therefore

$$F_{3\_5} = \frac{0.001 \times 2 \times 10^8 \times (-1.471\text{E-5})}{1.5} = \underline{-1.96 \text{ kN.}}$$

Member 4–5
From (3.29), $u_5 = 2.03\text{E-5}$
therefore

$$F_{4\_5} = \frac{0.001 \times 2 \times 10^8 \times 2.03\text{E-5}}{1.5} = \underline{2.71 \text{ kN.}}$$

3.6 COMPUTER PROGRAMS

Computer programs for the static analysis of pin-joined plane and space trusses appear in references [15] and [27], where the programs from the former are for a microcomputer and those from the latter are for a mainframe computer.

Using the 'truss' & 'space truss' programs of reference [15], the plane pin-jointed trusses of Example 3.2, together with the pin-jointed space truss of Example 3.3, were analysed on an Apple II computer, and Figs 3.15 and 3.16, show the deflected forms of two of these trusses. Both figures are of a qualitative nature, and in Fig. 3.16, the plan is shown on the left of the screen and the front elevation on the right.

EXAMPLES FOR PRACTICE

1. Determine the nodal displacements and member forces of the plane pin-jointed trusses shown in Figs 3.17 to 3.20. All members may be assumed to have a constant cross-sectional area 'A' and elastic modulus 'E', except for problems 1(b) and 1(d).

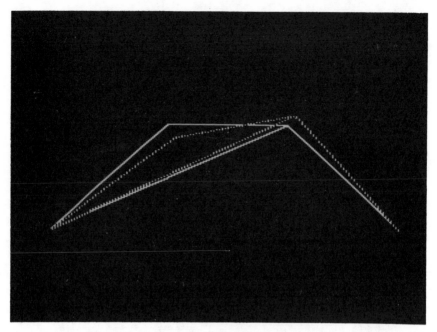

Fig. 3.15 – Deflected form of plane truss (example 3.2).

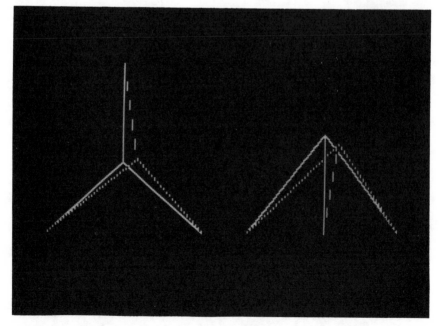

Fig. 3.16 – Deflected form of space truss (example 3.3). (Plan on left and front elevation on right)

(a)

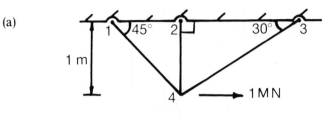

Fig. 3.17.

Answer

$$[\,u_4{}^\circ \quad v_4{}^\circ\,] = \frac{1}{AE}\,[\,1.397 \quad 0.129\,]$$

$$[\,F_{1\_4} \quad F_{2\_4} \quad F_{3\_4}\,] = [\,0.634 - 0.129 - 0.637\,]\,\text{MN}$$

(b)

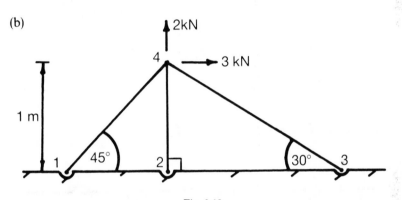

Fig. 3.18.

Member	A	E
1–4	$\sqrt{2A}$	E
2–4	A	E
3–4	$3A$	$3E$

Answer

$$[\,u_4{}^\circ \quad v_4{}^\circ\,] = \frac{1}{AE}\,[\,1.334 \quad 1.498\,]$$

$$[\,F_{1\_4} \quad F_{2\_4} \quad F_{3\_4}\,] = [\,2 \quad 1.5 \quad -1.829\,]\,\text{kN}.$$

(c)

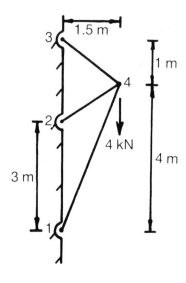

Fig. 3.19.

Answer

$$\lfloor u_4{}^\circ \quad v_4{}^\circ \rfloor = \frac{1}{AE} \lfloor 0.716 \quad -7.419 \rfloor$$

$$\lfloor F_{1\_4} \quad F_{2\_4} \quad F_{3\_4} \rfloor = \lfloor -1.57 \quad -1.95 \quad 2.61 \rfloor.$$

(d)

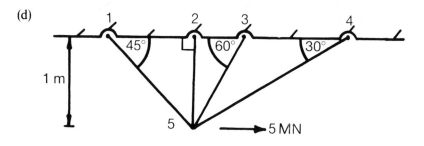

Fig. 3.20.

Member	A	E
1–5	2A	3E
2–5	A	E
3–5	A	E
4–5	3A	3E

Answer

$$[u_5^\circ \quad v_5^\circ] = \frac{1}{AE}[0.877 \quad -0.0362]$$

$$[F_{1-5} \quad F_{2-5} \quad F_{3-5} \quad F_{4-5}] = [2.738 \quad 0.0362 \quad -0.352 \quad -3.334] \text{ MN}.$$

2. Determine the nodal displacements and member forces of the pin-jointed space trusses (Figs 3.21, 3.22) of problems 2(a) and 2(b) respectively. The space truss of problem 2(a) is pinned firmly at nodes 1 to 4, and that of problem 2(b) at nodes 1 to 5.

(a)

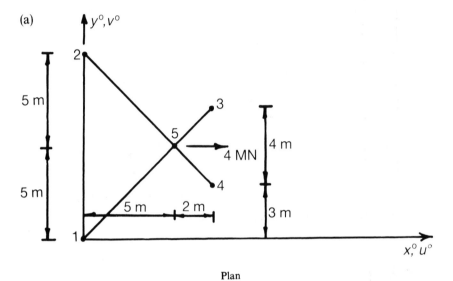

Plan

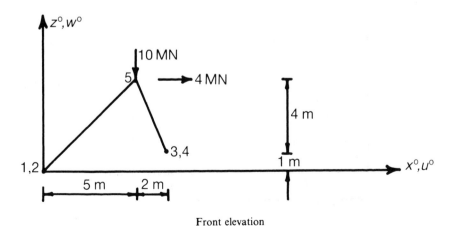

Front elevation

Fig. 3.21 – Plan front elevation.

All members may be assumed to have constant 'A' and 'E', where

A = cross-sectional area

E = elastic modulus.

Answer

$$\lfloor u_5^\circ \quad v_5^\circ \quad w_5^\circ \rfloor = \frac{1}{AE} \lfloor 17.09 \quad 41.37 \quad -25.75 \rfloor$$

$$\lfloor F_{1-5} \quad F_{2-5} \quad F_{3-5} \quad F_{4-5} \rfloor = \lfloor 2.18 \quad -3.34 \quad -9.16 \quad -2.27 \rfloor.$$

(b)

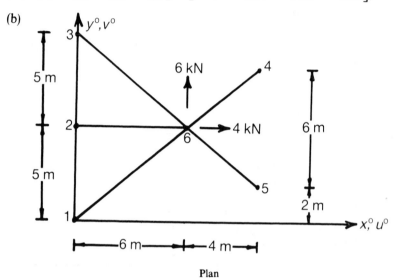

Plan

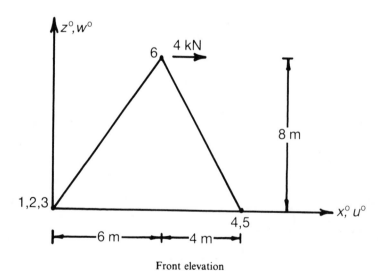

Front elevation

Fig. 3.22 – Plan front elevation.

Answer

$$\lfloor u_6{}^\circ \quad v_6{}^\circ \quad w_6{}^\circ \rfloor = \frac{1}{AE} \lfloor 33.25 \quad 104.9 \quad -4.37 \rfloor$$

$$\lfloor F_{1\_6} \quad F_{2\_6} \quad F_{3\_6} \quad F_{4\_6} \quad F_{5\_6} \rfloor = \lfloor 5.51 \quad 1.65 \quad -2.88 \\ -5.42 \quad 1.65 \ \rfloor \, kN.$$

Chapter 4

Static analysis of rigid-jointed frames

The static analysis of rigid-jointed frames and continuous beams is one of the most important aspects of structural engineering.

Rigid-jointed frames take many and varied forms, and appear in many different branches of engineering, varying from ships to aircraft and motor cars to off-shore drilling rigs.

Prior to the development of matrix methods in structural analysis, analytical methods were used, but, in general, these could only be applied to the simplest of structures. Even with two-dimensional rigid-jointed frames, the analysis became very tedious, particularly with sidesway problems in skew frames.

Today, however, with the vast army of available computers, these problems have become more of historic interest, than of analytical challenge. Even with only a microcomputer, the static analysis of continuous beams and rigid-jointed plane frames can be carried out quite successfully. For example, the static analysis of a 199-node continuous beam was successfully carried out by the author on a 32 k CMB PET in only about 8 minutes [15].

Complex shapes and boundary conditions, with complex combinations of load systems, no longer present a problem, and the numerical difficulties previously experienced with large numbers of redundancies, is a problem of the past.

There are, however, other problems that now occur, the most common being the shortage of computer memory, particularly for three-dimensional problems, and also the lack of precision due to numerical instability. Computational time can also present a problem.

In this chapter, the stiffness matrix of a beam is first developed, and later, this is extended to obtain the stiffness matrices of rigid-jointed frames. Applications are made to beams and two- and three-dimensional rigid-jointed frames.

4.1 DERIVATION OF SLOPE-DEFLECTION EQUATIONS

To obtain the elemental stiffness matrix for a beam, it will be found convenient to first derive the slope-deflection equations.

Consider the uniform straight beam of Fig. 4.1, which is subjected to clockwise couples M_1 and M_2 at its nodal points 1 and 2 respectively, together with the vertical forces Y_1 and Y_2.

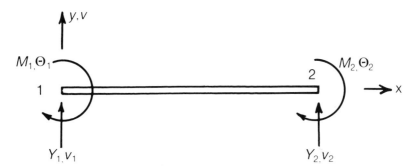

Fig. 4.1 – Beam element.

Assume the following:

v_1, v_2 = displacements in the 'y' direction at nodes 1 and 2, respectively.

θ_1, θ_2 = clockwise rotations of nodes 1 and 2, respectively.

Resolving vertically,

$$Y_1 = -Y_2 \tag{4.1}$$

Taking moments about node 2,

$$Y_1 = -\frac{(M_1 + M_2)}{l} \tag{4.2}$$

From (4.1),

$$Y_2 = \frac{(M_1 + M_2)}{l} \tag{4.3}$$

From small-deflection elastic theory,

$$EI\frac{d^2v}{dx^2} = M = Y_1 x + M_1.$$

On integrating

$$EI\frac{dv}{dx} = \frac{Y_1 x^2}{2} + M_1 x + A$$

and

$$EI v = \frac{Y_1 x^3}{6} + \frac{M_1 x^2}{2} + Ax + B$$

at

$$x = 0, \frac{dv}{dx} = -\theta_1, \text{ and } v = v_1,$$

therefore

$$A = -EI\theta_1 \qquad\qquad (4.4)$$

and

$$B = EIv_1 \qquad\qquad (4.5)$$

at

$$x = l, v = v_2 \text{ and } \frac{dv}{dx} = -\theta_2,$$

therefore

$$EI\theta_2 = \frac{Y_1 l^2}{2} + M_1 l - EI\theta_1 \qquad\qquad (4.6)$$

and

$$EI v_2 = \frac{Y_1 l^3}{6} + \frac{M_1 l^2}{2} - EI\theta_1 l + EI v_1. \qquad\qquad (4.7)$$

From (4.4) to (4.7),

$$M_1 = \frac{4EI\theta_1}{l} + \frac{2EI\theta_2}{l} - \frac{6EI}{l^2}(v_1 - v_2) \qquad\qquad (4.8)$$

$$Y_1 = -\frac{6EI\theta_1}{l^2} - \frac{6EI\theta_2}{l^2} + \frac{12EI}{l^3}(v_1 - v_2) \qquad\qquad (4.9)$$

$$M_2 = \frac{2EI\theta_1}{l} + \frac{4EI\theta_2}{l} - \frac{6EI}{l^2}(v_1 - v_2) \qquad\qquad (4.10)$$

$$Y_2 = \frac{6EI\,\theta_1}{l^2} + \frac{6EI\,\theta_2}{l^2} - \frac{12EI}{l^3}(v_1 - v_2). \tag{4.11}$$

Equations (4.8) to (4.11) are known as the *slope-deflection equations*, and they can be rearranged in the matrix form of (4.12):

$$\begin{Bmatrix} Y_1 \\ M_1 \\ Y_2 \\ M_2 \end{Bmatrix} = EI \begin{bmatrix} 12/l^3 & -6/l^2 & -12/l^3 & -6/l^2 \\ -6/l^2 & 4/l & 6/l^2 & 2/l \\ -12/l^3 & 6/l^2 & 12/l^3 & 6/l^2 \\ -6/l^2 & 2/l & 6/l^2 & 4/l \end{bmatrix} \begin{Bmatrix} v_1 \\ \theta_1 \\ v_2 \\ \theta_2 \end{Bmatrix}.$$

Equation (4.12) can be put in the form:

$$\{P_i\} = [k]\{u_i\}$$

where

$$\{P_i\} = \begin{Bmatrix} Y_1 \\ M_1 \\ Y_2 \\ M_2 \end{Bmatrix} = \text{a vector of generalised loads}$$

$$\{u_i\} = \begin{Bmatrix} v_1 \\ \theta_1 \\ v_2 \\ \theta_2 \end{Bmatrix} = \text{a vector of generalised displacements}$$

$[k]$ = the elemental stiffness matrix for a beam

$$= EI \begin{bmatrix} \overset{v_1}{12/l^3} & \overset{\theta_1}{-6/l^2} & \overset{v_2}{-12/l^3} & \overset{\theta_2}{-6/l^2} \\ -6/l^2 & 4/l & 6/l^2 & 2/l \\ -12/l^3 & 6/l^2 & 12/l^3 & 6/l^2 \\ -6/l^2 & 2/l & 6/l^2 & 4/l \end{bmatrix} \begin{matrix} v_1 \\ \theta_1 \\ v_2 \\ \theta_2 \end{matrix} \tag{4.13}$$

4.2 SOLUTION OF BEAMS

Solution of beams can be carried out in a manner similar to that of Chapter 3, except that the elemental stiffness matrix now adopted is (4.13), and the vector of loads contains a rotational load M, and that this couple corresponds to a rotational displacement θ.

    To demonstrate the method, the following example will be considered.

4.2.1   Example 4.1

Determine the nodal displacements and moments for the uniform section encastré beam of Fig. 4.2.

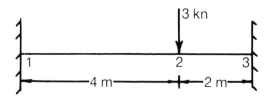

Fig. 4.2 – Encastré beam.

Element 1–2

$$l = 4m.$$

Substituting this value of l into equation (4.13), the following is obtained for the elemental stiffness matrix for element 1–2.

$$
[k_{1-2}] = EI
\begin{array}{cccc}
v_1 & \theta_1 & v_2 & \theta_2 \\
\end{array}
\left[
\begin{array}{cccc}
12/64 & -6/16 & -12/64 & -6/16 \\
-6/16 & 4/4 & 6/16 & 2/4 \\
-12/64 & 6/16 & 12/64 & 6/16 \\
-6/16 & 2/4 & 6/16 & 4/4
\end{array}
\right]
\begin{array}{c}
v_1 \\
\theta_1 \\
v_2 \\
\theta_2.
\end{array}
\tag{4.14}
$$

Element 2–3

$$l = 2m.$$

Substituting this value of l into (4.13), the following is obtained for the elemental stiffness matrix for element 2–3.

$$
[k_{2-3}] = EI
\begin{array}{cccc}
v_2 & \theta_2 & v_3 & \theta_3 \\
\end{array}
\left[
\begin{array}{cccc}
12/8 & -6/4 & -12/8 & -6/4 \\
-6/4 & 4/2 & 6/4 & 2/2 \\
-12/8 & 6/4 & 12/8 & 6/4 \\
-6/4 & 2/2 & 6/4 & 4/2
\end{array}
\right]
\begin{array}{c}
v_2 \\
\theta_2 \\
v_3 \\
\theta_3.
\end{array}
\tag{4.15}
$$

To obtain $[K_{11}]$, the structural stiffness matrix corresponding to the free displacements, add together those parts of the stiffness influence coefficients of (4.14) and (4.15), corresponding to the free displacement v_2 and θ_2, as in (4.16).

$$
[K_{11}] = EI
\begin{array}{cc}
v_2 & \theta_2 \\
\end{array}
\left[
\begin{array}{cc}
12/64 + 12/8 & 6/16 - 6/4 \\
6/16 - 6/4 & 4/4 + 4/2
\end{array}
\right]
\begin{array}{c}
v_2 \\
\theta_2.
\end{array}
\tag{4.16}
$$

The load vector corresponding to the free displacements v_2 and θ_2, is

$$\{q_F\} = \begin{Bmatrix} -3 \\ 0 \end{Bmatrix} \begin{matrix} v_2 \\ \theta_2 \end{matrix}$$

which results in the following set of simultaneous equations:

$$\begin{Bmatrix} -3 \\ 0 \end{Bmatrix} = EI \begin{bmatrix} 1.688 & -1.125 \\ -1.125 & 3 \end{bmatrix} \begin{Bmatrix} v_2 \\ \theta_2 \end{Bmatrix}$$

or

$$\begin{Bmatrix} v_2 \\ \theta_2 \end{Bmatrix} = \frac{1}{EI} \frac{\begin{bmatrix} 3 & 1.125 \\ 1.125 & 1.688 \end{bmatrix}}{3.798} \begin{Bmatrix} -3 \\ 0 \end{Bmatrix}$$

$$= \frac{1}{EI} \begin{Bmatrix} -2.37 \\ -0.889 \end{Bmatrix}.$$

The negative values for v_2 and θ_2 indicate that at node 2, the vertical deflection is downward and the rotation is anticlockwise.

To obtain the nodal moments M_1, M_2 and M_3, substitute the values for nodal displacements into the slope-deflection equations (4.8) and (4.10), as follows.

Element 1–2

$$M_1 = \frac{4EI}{4}0 + \frac{2EI}{4}\left(\frac{-0.889}{EI}\right) - \frac{6EI}{4^2}\frac{1}{EI}\{0-(-2.37)\}$$

$$M_1 = -1.333 \text{ kN m}$$

$$M_2 = \frac{2EI}{4}0 + \frac{4EI}{4}\left(\frac{-0.889}{EI}\right) - \frac{6EI}{4^2}\frac{1}{EI}\{0-(-2.37)\}$$

$$M_2 = -1.778 \text{ kN m}$$

Element 2–3

$$M_2 = \frac{4EI}{2}\left(\frac{-0.889}{EI}\right) - \frac{6EI}{2^2}\frac{1}{EI}(-2.37-0)$$

$$M_2 = 1.777 \text{ kN m}$$

$$M_3 = 2.666 \text{ kN m}$$

N.B. It should be noted that the moments either side of the node should be in equilibrium, as they are in this case for M_2.

4.2.2   Distributed loading

To consider the effects of distributed loading, the examples of Figs 4.3 and 4.4 will be considered.

4.2.3   Example 4.2

Calculate the moments at the supports of the beam of Fig. 4.3, which may be assumed to be fixed at node 1 and supported at the same level at nodes 2 and 3.

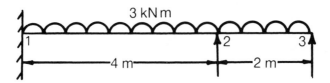

Fig. 4.3 – Continuous beam.

The following method is considered the most simple for distributed loads on beams and frames:

(a) Consider the beam to be encastré at the supports (which are taken as the nodal points), and calculate the end fixing forces, as shown in Fig. 4.4(a). (For the present problem, only bending moments will be considered, as the supports are assumed rigid in the vertical direction).

(b) The beam in condition (a) is not in equilibrium at the nodal points, hence, the negative resultants of the end fixing forces must be applied at these points to achieve equilibrium, as shown in Fig. 4.4(b).

(c) Application of the negative resultants of the end fixing forces to the beam will cause it to deform at the nodes, and the internal forces due to this must be calculated by the displacement method.

(d) Superposition of the moments calculated from (a) with those calculated from (c) will give the actual moments of the structure at its nodal points due to the loading condition of Fig. 4.3.

There is no resultant moment at node 1, because it is fixed at this node and the moment of 4 kN m is absorbed by the wall. Similarly, there are no vertical resultant forces, as all the nodes have rigid supports in the vertical direction.

There are two degrees of freedom, namely θ_2 and θ_3, and the vector of nodal forces corresponding to these is $\{q_F\}$, Fig. 4.4(b), which is given by

$$q_F = \begin{Bmatrix} -3 \\ -1 \end{Bmatrix} \text{kN m} \tag{4.17}$$

The negative values in $\{q_F\}$ are because the external nodal moments are anticlockwise.

The solution now follows a pattern similar to that adopted for the

(a) End fixing moments

4 kN m   4 kN m

1 kN m   1 kN m

(b) Negative resultants of end moments

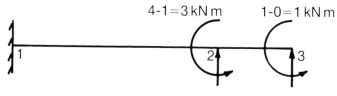

Fig. 4.4.

example of Fig. 4.2, except that the end fixing moments of Fig. 4.3(a) have to be superimposed with those moments obtained due to the application of $\{q_F\}$.

Member 1–2

$l = 4$ m.

Substituting this value into (4.13), the elemental stiffness matrix for element 1–2 is given by (4.18):

$$[k_{1-2}] = EI \begin{matrix} & v_1 & \theta_1 & v_2 & \theta_2 & \\ \begin{bmatrix} 12/64 & -6/16 & -12/64 & -6/16 \\ -6/16 & 4/4 & 6/16 & 2/4 \\ -12/64 & 6/16 & 12/64 & 6/16 \\ -6/16 & 2/4 & 6/16 & 4/4 \end{bmatrix} & \begin{matrix} v_1 \\ \theta_1 \\ v_2 \\ \theta_2 \, . \end{matrix} \end{matrix} \quad (4.18)$$

Member 2–3

$l = 2$ m.

Substituting this value into (4.13),

$$[k_{2-3}] = EI \begin{array}{cccc} v_2 & \theta_2 & v_3 & \theta_3 \\ \begin{bmatrix} 12/8 & -6/4 & -12/8 & -6/4 \\ -6/4 & 4/2 & 6/4 & 2/2 \\ -12/8 & 6/4 & 12/8 & 6/4 \\ -6/4 & 2/2 & 6/4 & 4/2 \end{bmatrix} & \begin{array}{c} v_2 \\ \theta_2 \\ v_3 \\ \theta_3. \end{array} \end{array} \quad (4.19)$$

To obtain $[K_{11}]$, the stiffness influence coefficients of (4.18) and (4.19), corresponding to the free displacements θ_2 and θ_3, must be added together in (4.20):

$$[K_{11}] = EI \begin{array}{cc} \theta_2 & \theta_3 \\ \begin{bmatrix} 4/4 + 4/2 & 2/2 \\ 2/2 & 4/2 \end{bmatrix} & \begin{array}{c} \theta_2 \\ \theta_3. \end{array} \end{array} \quad (4.20)$$

From (4.17) and (4.20), the resulting simultaneous equations are

$$\begin{Bmatrix} -3 \\ -1 \end{Bmatrix} = EI \begin{bmatrix} 3 & 1 \\ 1 & 2 \end{bmatrix} \begin{Bmatrix} \theta_2 \\ \theta_3 \end{Bmatrix}.$$

Hence

$$\begin{Bmatrix} \theta_2 \\ \theta_3 \end{Bmatrix} = \frac{1}{EI} \frac{\begin{bmatrix} 2 & -1 \\ -1 & 3 \end{bmatrix}}{5} \begin{Bmatrix} -3 \\ -1 \end{Bmatrix}$$

$$= \frac{1}{EI} \begin{Bmatrix} -1 \\ 0 \end{Bmatrix}.$$

To calculate the nodal moments, M_1, M_2, *and* M_3

Element 1–2
From the slope-deflection equations (4.8) and (4.10),

$$\begin{Bmatrix} M_1 \\ M_2 \end{Bmatrix} = EI \begin{bmatrix} -0.375 & 1 & 0.375 & 0.5 \\ -0.375 & 0.5 & 0.375 & 1 \end{bmatrix} \begin{Bmatrix} v_1 \\ \theta_1 \\ v_2 \\ \theta_2 \end{Bmatrix} + \begin{Bmatrix} \text{End fixing} \\ \text{moments} \end{Bmatrix}$$

$$= EI \begin{bmatrix} -0.375 & 1 & 0.375 & 0.5 \\ -0.375 & 0.5 & 0.375 & 1 \end{bmatrix} \begin{Bmatrix} 0 \\ 0 \\ 0 \\ -1/EI \end{Bmatrix} + \begin{Bmatrix} -4 \\ 4 \end{Bmatrix}$$

$$\begin{Bmatrix} M_1 \\ M_2 \end{Bmatrix} = \begin{Bmatrix} -4.5 \\ 3 \end{Bmatrix} \text{ kN m.}$$

Element 2–3
Similarly,

$$\begin{Bmatrix} M_2 \\ M_3 \end{Bmatrix} = EI \begin{bmatrix} -1.5 & 2 & 1.5 & 1 \\ -1.5 & 1 & 1.5 & 2 \end{bmatrix} \begin{Bmatrix} v_2 \\ \theta_2 \\ v_3 \\ \theta_3 \end{Bmatrix} + \begin{Bmatrix} \text{End fixing} \\ \text{moments} \end{Bmatrix}$$

$$= EI \begin{bmatrix} -1.5 & 2 & 1.5 & 1 \\ -1.5 & 1 & 1.5 & 2 \end{bmatrix} \begin{Bmatrix} 0 \\ -1/EI \\ 0 \\ 0 \end{Bmatrix} + \begin{Bmatrix} -1 \\ 1 \end{Bmatrix}$$

$$\begin{Bmatrix} M_2 \\ M_3 \end{Bmatrix} = \begin{Bmatrix} -3 \\ 0 \end{Bmatrix} \text{kN m.}$$

4.2.4   Example 4.3

Determine the nodal displacements and moments for the continuous beam of
Fig. 4.5.

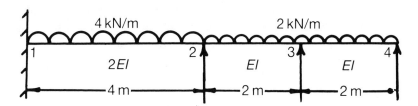

Fig. 4.5 – Varying section continuous beam.

    The end fixing forces are shown for each element in Fig. 4.6(a), and the
negative resultants at the nodes are shown in Fig. 4.6(b).
    From Fig. 4.6(b), it can be seen that the vector of external nodal loads is
given by (4.21):

$$\{q_F\} = \begin{Bmatrix} -4.667 \\ 0 \\ -0.667 \end{Bmatrix} \text{kN m.} \tag{4.21}$$

(a) *End fixing moments for elements*

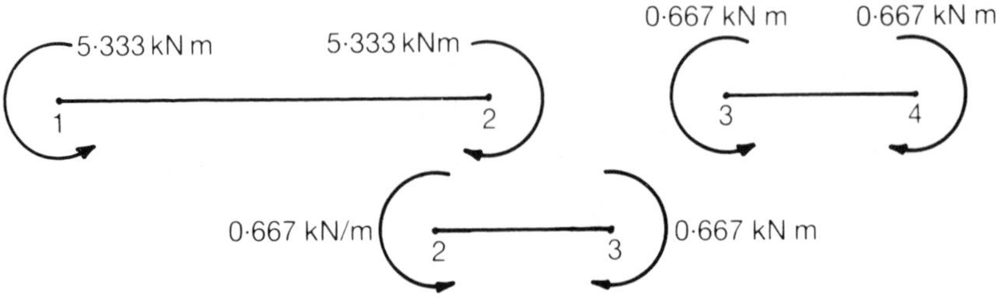

(b) *Negative nodal resultants of (a)*

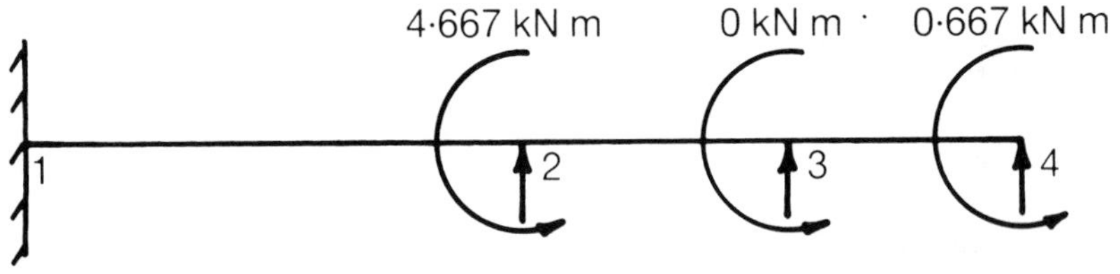

Fig. 4.6.

There are three degrees of freedom, namely θ_2, θ_3, and θ_4, and these correspond to the external nodal couples of $\{q_F\}$ (See Fig. 4.6(b)).

As there are only three degrees of freedom, the elemental stiffness matrices can be determined by removing all columns and rows, which do not correspond to these three free displacements, as follows.

Member 1–2

$l = 4$ m.

$$[k_{1-2}] = 2EI \overset{\theta_2}{[4/4]}\theta_2 = EI \overset{\theta_2}{[2]}\theta_2. \tag{4.22}$$

Member 2–3

$l = 2$ m

$$[k_{2-3}] = EI \quad \begin{array}{c} \begin{matrix} \theta_2 & \theta_3 \end{matrix} \\ \begin{bmatrix} 4/2 & 2/2 \\ 2/2 & 4/2 \end{bmatrix} \end{array} \begin{matrix} \theta_2 \\ \theta_3 \end{matrix} \tag{4.23}$$

Member 3–4

$l = 2$ m.

$$[k_{3-4}] = EI \quad \begin{matrix} \theta_3 & \theta_4 \\ \begin{bmatrix} 2 & 1 \\ 1 & 2 \end{bmatrix} & \begin{matrix} \theta_3 \\ \theta_4. \end{matrix} \end{matrix} \tag{4.24}$$

$[K_{11}]$ is obtained by adding together the appropriate stiffness influence coefficients of (4.22) to (4.24), as in (4.25):

$$[K_{11}] = EI \quad \begin{matrix} \theta_2 & \theta_3 & \theta_4 \\ \begin{bmatrix} 2+2 & 1 & \\ 1 & 2+2 & 1 \\ & 1 & 2 \end{bmatrix} & \begin{matrix} \theta_2 \\ \theta_3 \\ \theta_4. \end{matrix} \end{matrix} \tag{4.25}$$

From (4.21) and (4.25), the relationship between forces and displacements is given by:

$$\begin{Bmatrix} -4.667 \\ 0 \\ -0.667 \end{Bmatrix} = EI \begin{bmatrix} 4 & 1 & 0 \\ 1 & 4 & 1 \\ 0 & 1 & 2 \end{bmatrix} \begin{Bmatrix} \theta_2 \\ \theta_3 \\ \theta_4 \end{Bmatrix}.$$

Solution of the above results in the following:

$$\begin{Bmatrix} \theta_2 \\ \theta_3 \\ \theta_4 \end{Bmatrix} = \frac{1}{EI} \begin{Bmatrix} -1.282 \\ 0.462 \\ -0.564 \end{Bmatrix}.$$

To calculate the nodal moments

Member 1–2
From the slope-deflection equations (4.8) and (4.10),

$$\begin{Bmatrix} M_1 \\ M_2 \end{Bmatrix} = EI \begin{bmatrix} -0.75 & 2 & 0.75 & 1 \\ -0.75 & 1 & 0.75 & 2 \end{bmatrix} \frac{1}{EI} \begin{Bmatrix} v_1 \\ \theta_1 \\ v_2 \\ \theta_2 \end{Bmatrix} + \begin{Bmatrix} \text{End fixing} \\ \text{moments} \end{Bmatrix}$$

$$\begin{Bmatrix} M_1 \\ M_2 \end{Bmatrix} = EI \begin{bmatrix} -0.75 & 2 & 0.75 & 1 \\ -0.75 & 1 & 0.75 & 2 \end{bmatrix} \frac{1}{EI} \begin{Bmatrix} 0 \\ 0 \\ 0 \\ -1.282 \end{Bmatrix} + \begin{Bmatrix} -5.333 \\ 5.333 \end{Bmatrix}$$

$$= \begin{Bmatrix} -6.615 \\ 2.769 \end{Bmatrix} \text{ kN m.}$$

Member 2–3
From the slope–deflection equations

$$\begin{Bmatrix} M_2 \\ M_3 \end{Bmatrix} = EI \begin{bmatrix} -1.5 & 2 & 1.5 & 1 \\ -1.5 & 1 & 1.5 & 2 \end{bmatrix} \frac{1}{EI} \begin{Bmatrix} v_2 \\ \theta_2 \\ v_3 \\ \theta_3 \end{Bmatrix} + \begin{Bmatrix} \text{End fixing} \\ \text{moments} \end{Bmatrix}$$

$$= EI \begin{bmatrix} -1.5 & 2 & 1.5 & 1 \\ -1.5 & 1 & 1.5 & 2 \end{bmatrix} \frac{1}{EI} \begin{Bmatrix} 0 \\ -1.282 \\ 0 \\ -0.462 \end{Bmatrix} + \begin{Bmatrix} -0.6667 \\ 0.6667 \end{Bmatrix}$$

$$= \begin{Bmatrix} -2.769 \\ 0.308 \end{Bmatrix} \text{ kN m.}$$

Member 3–4

$$\begin{Bmatrix} M_3 \\ M_4 \end{Bmatrix} = EI \begin{bmatrix} -1.5 & 2 & 1.5 & 1 \\ -1.5 & 1 & 1.5 & 2 \end{bmatrix} \frac{1}{EI} \begin{Bmatrix} 0 \\ 0.462 \\ 0 \\ -0.564 \end{Bmatrix} + \begin{Bmatrix} -0.6667 \\ 0.6667 \end{Bmatrix}$$

$$= \begin{Bmatrix} -0.307 \\ 0 \end{Bmatrix} \text{ kN m.}$$

N.B. All elements for horizontal beams should be defined by their low node first and their high node second.

4.3 RIGID-JOINTED PLANE FRAMES

The elemental stiffness matrix for a beam, namely equation (4.13), can be extended for the analysis of rigid-jointed plane frames, providing it is modified to allow for the element to be inclined at an angle to the horizontal, and also, if the stiffness matrix for a two-dimensional rod is superimposed with this matrix.

Consider first a one-dimensional element inclined at an angle 'α' to the x^o axis, as shown in Fig. 4.7, where,

X = force in x direction

Y = force in y direction

X^o = force in x^o direction

Y^o = force in y^o direction

M = moment in x–y or x^o–y^o planes

x–y = local axes

x^o–y^o = global axes

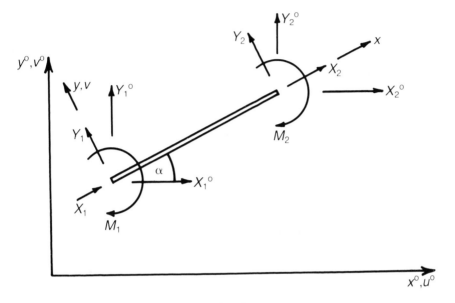

Fig. 4.7.

To represent the force–displacement relationships of the element of Fig. 4.7 in local coordinates, the slope–deflection equations of (4.8) to (4.11) are combined with the force–deflection equations of (3.1) and (3.2), as follows.

$$
\begin{Bmatrix} X_1 \\ Y_1 \\ M_1 \\ X_2 \\ Y_2 \\ M_2 \end{Bmatrix} = EI
\begin{bmatrix}
(A/lI) & 0 & 0 & (-A/lI) & 0 & 0 \\
0 & 12/l^3 & -6/l^2 & 0 & -12/l^3 & -6/l^2 \\
0 & -6/l^2 & 4/l & 0 & 6/l^2 & 2/l \\
(-A/lI) & 0 & 0 & (A/lI) & 0 & 0 \\
0 & -12/l^3 & 6/l^2 & 0 & 12/l^3 & 6/l^2 \\
0 & -6/l^2 & 2/l & 0 & 6/l^2 & 4/l
\end{bmatrix}
\begin{Bmatrix} u_1 \\ v_1 \\ \theta_1 \\ u_2 \\ v_2 \\ \theta_2 \end{Bmatrix}
$$

$$(4.26)$$

or

$$\{p_i\} = [k]\{u_i\}.$$

The relationship between local and global forces is clearly given by:

$$
\begin{Bmatrix} X_1 \\ Y_1 \\ M_1 \\ X_2 \\ Y_2 \\ M_2 \end{Bmatrix}
=
\left[
\begin{array}{ccc|ccc}
c & s & 0 & & & \\
-s & c & 0 & & 0_3 & \\
0 & 0 & 1 & & & \\
\hline
& & & c & s & 0 \\
& 0_3 & & -s & c & 0 \\
& & & 0 & 0 & 1
\end{array}
\right]
\begin{Bmatrix} X_1{}^{\circ} \\ Y_1{}^{\circ} \\ M_1 \\ X_2{}^{\circ} \\ Y_2{}^{\circ} \\ M_2 \end{Bmatrix}
\qquad (4.27)
$$

or

$$\{p_i\} = [\Xi]\,\{p_i{}^{\circ}\}$$

where

$$c = \cos\alpha;\ s = \sin\alpha.$$

Substituting $[k]$ from (4.26) and $[\Xi]$ from (4.27) into (3.14), the elemental stiffness matrix for the general case of a plane one-dimensional element, in global coordinates, is given by

$$[k^{\circ}] = [k_b{}^{\circ}] + [k_r{}^{\circ}]$$

where

$$
[k_b{}^{\circ}] = EI
\left[
\begin{array}{cccccc}
\frac{12}{l^3}s^2 & & & & & \\[2mm]
-\frac{12}{l^3}cs & \frac{12}{l^3}c^2 & & \text{Symmetrical} & & \\[2mm]
\frac{6}{l^2}s & -\frac{6}{l^2}c & \frac{4}{l} & & & \\[2mm]
\hline
-\frac{12}{l^3}s^2 & \frac{12}{l^3}cs & -\frac{6}{l^2}s & \frac{12}{l^3}s^2 & & \\[2mm]
\frac{12}{l^3}cs & -\frac{12}{l^3}c^2 & \frac{6}{l^2}c & -\frac{12}{l^3}cs & \frac{12}{l^3}c^2 & \\[2mm]
\frac{6}{l^2}s & -\frac{6}{l}c & \frac{2}{l} & -\frac{6}{l^2}s & \frac{6}{l^2}c & \frac{4}{l}
\end{array}
\right]
\begin{array}{l} u_1{}^{\circ} \\[2mm] v_1{}^{\circ} \\[2mm] \theta_1 \\[2mm] u_2{}^{\circ} \\[2mm] v_2{}^{\circ} \\[2mm] \theta_2 \end{array}
$$

with column headings $u_1{}^{\circ}\ \ v_1{}^{\circ}\ \ \theta_1\ \ u_2{}^{\circ}\ \ v_2{}^{\circ}\ \ \theta_2$

$$\qquad (4.28)$$

= stiffness matrix for an inclined beam element.

$$
\mathrm{a} \qquad [k_r^\circ] = \frac{AE}{l}
\begin{array}{c}
\begin{array}{cccccc}
u_1^\circ & v_1^\circ & \theta_1 & u_2^\circ & v_2^\circ & \theta
\end{array} \\
\begin{bmatrix}
c^2 & cs & 0 & -c^2 & -cs & 0 \\
cs & s^2 & 0 & -cs & -s^2 & 0 \\
0 & 0 & 0 & 0 & 0 & 0 \\
-c^2 & -cs & 0 & c^2 & cs & 0 \\
-cs & -s^2 & 0 & cs & s^2 & 0 \\
0 & 0 & 0 & 0 & 0 & 0
\end{bmatrix}
\begin{array}{c}
u_1^\circ \\ v_1^\circ \\ \theta_1 \\ u_2^\circ \\ v_2^\circ \\ \theta_2
\end{array}
\end{array}
\qquad (4.29)
$$

= stiffness matrix for an inclined rod element.

To demonstrate the application of (4.28) and (4.29) to rigid-jointed plane frames, the frameworks of Fig. 4.8 and 4.9 will be considered.

4.3.1   Example 4.4

Determine the nodal displacements and forces for the constant section rigid-jointed plane frame of Fig. 4.8. The framework is firmly fixed at nodes 1 and 4, and to simplify the solution, it may be assumed that the axial stiffnesses of the elements is large when compared to their flexural stiffnesses, so that it is reasonable to assume that

$$v_2^\circ = v_3^\circ = 0;\ u_2^\circ = u_3^\circ;\quad \text{and}\quad \theta_2 = \theta_3.$$

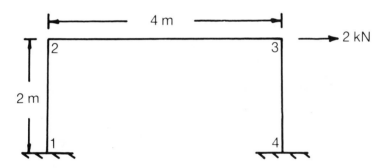

Fig. 4.8 – Portal frame.

As there are only four degrees of freedom, namely, u_2°, θ_2, u_3°, and θ_3, it will be found convenient to eliminate all columns and rows of the elemental stiffness matrices, corresponding to the zero displacements. This process will considerably simplify computation.

Member 1–2

$\alpha = 90^\circ.$

$c = 0;\ s = 1;\ l = 2\text{m}.$

Substituting these values into (4.28) and (4.29), and removing all rows and columns corresponding to zero displacements, the elemental stiffness matrix for this element, in global coordinates, is given by (4.30):

$$[k_{1-2}^{\,\circ}] = EI \quad \begin{matrix} u_2^{\,\circ} & \theta_2 \\ \begin{bmatrix} 1.5 & -1.5 \\ -1.5 & 2 \end{bmatrix} & \begin{matrix} u_2^{\,\circ} \\ \theta_2. \end{matrix} \end{matrix} \qquad (4.30)$$

Member 2–3

$\alpha = 0; c = 1; s = 0; l = 4\,\mathrm{m}.$

Substituting these values into (4.28) and (4.29), and removing all rows and columns corresponding to zero displacements, the elemental stiffness matrix for element 2–3, in global coordinates, is given by (4.31):

$$[k_{2-3}^{\,\circ}] = EI \begin{matrix} \theta_2 & \theta_3 \\ \begin{bmatrix} 1 & 0.5 \\ 0.5 & 1 \end{bmatrix} & \begin{matrix} \theta_2 \\ \theta_3. \end{matrix} \end{matrix} \qquad (4.31)$$

Member 3–4

$\alpha = 270^\circ; c = 0; s = -1.$

$$[k_{3-4}^{\,\circ}] = EI \begin{matrix} u_3^{\,\circ} & \theta_3 \\ \begin{bmatrix} 1.5 & -1.5 \\ -1.5 & 2 \end{bmatrix} & \begin{matrix} u_3^{\,\circ} \\ \theta_3. \end{matrix} \end{matrix} \qquad (4.32)$$

To obtain $[K_{11}]$, the appropriate parts of the stiffness influence coefficients of (4.30) to (4.32) are added together, as in (4.33):

$$[K_{11}] = EI \begin{matrix} u_2^{\,\circ} & \theta_2 & u_3^{\,\circ} & \theta_3 \\ \begin{bmatrix} 1.5 & -1.5 & 0 & 0 \\ -1.5 & 3 & 0 & 0.5 \\ 0 & 0 & 1.5 & -1.5 \\ 0 & 0.5 & -1.5 & 3 \end{bmatrix} & \begin{matrix} u_2^{\,\circ} \\ \theta_2 \\ u_3^{\,\circ} \\ \theta_3 \,. \end{matrix} \end{matrix} \qquad (4.33)$$

Rewriting these as simultaneous equations, and bearing in mind that as $u_2^{\,\circ} = u_3^{\,\circ}$, the 2 kN load must be shared equally between nodes 2 and 3, the following is obtained:

$1 = EI(1.5u_2^{\,\circ} - 1.5\theta_2)$

$0 = EI(-1.5u_2^{\,\circ} + 3\theta_2 + 0.5\theta_3)$

$1 = EI(1.5u_3^{\,\circ} - 1.5\theta_3)$

$0 = EI(0.5\theta_2 - 1.5u_3^{\,\circ} + 3\theta_3),$

but $u_2{}^\circ = u_3{}^\circ$ and $\theta_2 = \theta_3$, therefore the above four simultaneous equations reduce to the following two simultaneous equations:

$$1 = EI(1.5u_2{}^\circ - 1.5\theta_2)$$
$$0 = EI(-1.5u_2{}^\circ + 3.5\theta_2).$$

Solving these two simultaneous equations, results in the following:

$$\theta_2 = \theta_3 = 0.5/EI$$
$$u_2{}^\circ = u_3{}^\circ = 1.16667/EI.$$

To determine the nodal moments, the local displacements of each element must first be calculated, and these must be substituted into the slope-deflection equations of a beam element, as follows:

Member 1–2

$$c = 0; s = 1.$$

$$\begin{Bmatrix} u_2 \\ v_2 \end{Bmatrix} = \begin{bmatrix} c & s \\ -s & c \end{bmatrix} \begin{Bmatrix} u_2{}^\circ \\ v_2{}^\circ \end{Bmatrix}$$

$$= \begin{bmatrix} 0 & 1 \\ -1 & 0 \end{bmatrix} \begin{Bmatrix} 1.667/EI \\ 0 \end{Bmatrix}$$

$$v_2 = -1.667/EI.$$

From (4.8) and (4.10),

$$\begin{Bmatrix} M_1 \\ M_2 \end{Bmatrix} = EI \begin{bmatrix} -6/2^2 & 4/2 & 6/2^2 & 2/2 \\ -6/2^2 & 2/2 & 6/2^2 & 4/2 \end{bmatrix} \begin{Bmatrix} v_1 \\ \theta_1 \\ v_2 \\ \theta_2 \end{Bmatrix}$$

$$= \begin{Bmatrix} 1.25 \\ -0.75 \end{Bmatrix} \text{kN m.}$$

Member 2–3

$$c = 1; s = 0$$

$$v_2 = \lfloor -s \quad c \rfloor \begin{Bmatrix} u_2{}^\circ \\ v_2{}^\circ \end{Bmatrix} = 0$$

$$v_3 = \lfloor -s \quad c \rfloor \begin{Bmatrix} u_3{}^\circ \\ v_3{}^\circ \end{Bmatrix} = 0.$$

From (4.8) and (4.10),

$$
\begin{Bmatrix} M_2 \\ M_3 \end{Bmatrix} = EI \begin{bmatrix} -6/4^2 & 4/4 & 6/4^2 & 2/4 \\ -6/4^2 & 2/4 & 6/4^2 & 4/4 \end{bmatrix} \begin{Bmatrix} v_2 \\ \theta_2 \\ v_3 \\ \theta_3 \end{Bmatrix}
$$

$$
= \begin{Bmatrix} 0.75 \\ 0.75 \end{Bmatrix} \text{kN m.}
$$

Member 3–4

$$c = 0; \quad s = -1$$

$$v_3 = \lfloor -s \quad c \rfloor \begin{Bmatrix} u_2{}^\circ \\ v_2{}^\circ \end{Bmatrix} = 1.667/EI.$$

From (4.8) and (4.10),

$$
\begin{Bmatrix} M_3 \\ M_4 \end{Bmatrix} = EI \begin{bmatrix} -1.5 & 2 & 1.5 & 1 \\ -1.5 & 1 & 1.5 & 2 \end{bmatrix} \begin{Bmatrix} 1.16667/EI \\ 0.5/EI \\ 0 \\ 0 \end{Bmatrix}
$$

$$
= \begin{Bmatrix} -0.75 \\ -1.25 \end{Bmatrix} \text{kN m.}
$$

N.B. It is not necessary to define the element of a rigid-jointed frame by its low node first and its high node second (see section 3).

4.3.2 Example 4.5

Determine the nodal displacements and moments for the rigid-jointed plane frame shown in Fig. 4.9. It may be assumed that the axial stiffnesses of the members is very large compared with their flexural stiffnesses, so that

$$v_2{}^\circ = v_3{}^\circ = 0 \quad \text{and} \quad u_2{}^\circ = u_3{}^\circ.$$

It should be noted that $\theta_2 \neq \theta_3$, because of the effects of the distributed loads on members 1–2 and 2–3.

To determine $\{q_F\}$

There are four degrees of freedom, namely, $u_2{}^\circ$, θ_2, $u_3{}^\circ$, and θ_3, and therefore $\{q_F\}$ will be of order 4×1, each coefficient of $\{q_F\}$ corresponding to a free displacement.

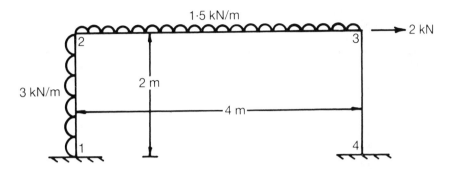

Fig. 4.9.

(a) To determine the effects of the distributed loads, it will first be necessary that each element is fixed at its nodes, and then to calculate the resulting end fixing forces corresponding to the free displacements, as in Fig. 4.10.

(b) Next, it will be necessary to determine the negative resultants of the end fixing forces, as shown in Fig. 4.11, and also to add the horizontal concentrated load of 2 kN at node 3.

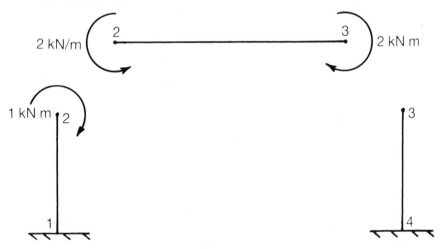

Fig. 4.10 – End fixing forces.

(c) As the member 2–3 has a very large axial stiffness, $u_2{}^\circ = u_3{}^\circ$, the horizontal loads at nodes 2 and 3 will be resisted by members 1–2 and 2–3 together. Later, it will be necessary to remember this, but for the time being, the vector $\{q_F\}$ can be taken as (4.34):

$$\{q_F\} \quad \begin{Bmatrix} 3 \\ 1 \\ 2 \\ -2 \end{Bmatrix} \quad \begin{matrix} u_2{}^\circ \\ \theta_2 \\ u_3{}^\circ \\ \theta_3. \end{matrix} \qquad (4.34)$$

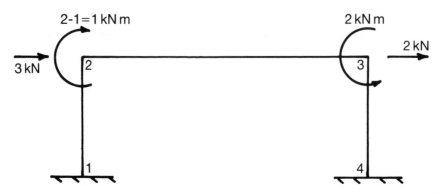

Fig. 4.11 – Negative resultants $+2$ kN load.

The elemental stiffness matrices are as in the previous problem, so

$$[K_{11}] = EI \begin{array}{cccc} u_2{}^\circ & \theta_2 & u_3{}^\circ & \theta_3 \\ \begin{bmatrix} 1.5 & -1.5 & 0 & 0 \\ -1.5 & 3 & 0 & 0.5 \\ 0 & 0 & 1.5 & -1.5 \\ 0 & 0.5 & -1.5 & 3 \end{bmatrix} & \begin{array}{c} u_2{}^\circ \\ \theta_2 \\ u_3{}^\circ \\ \theta_3 \,. \end{array} \end{array} \qquad (4.35)$$

Rewriting (4.34) and (4.35) in the form of simultaneous equations, and setting $u_3{}^\circ$ as $u_2{}^\circ$, the following is obtained:

$$\begin{aligned}
3 &= EI(1.5u_2{}^\circ - 1.5\theta_2) & \text{(a)} \\
1 &= EI(-1.5u_2{}^\circ + 3\theta_2 + 0.5\theta_3) & \text{(b)} \\
2 &= EI(1.5u_2{}^\circ - 1.5\theta_3) & \text{(c)} \\
-2 &= EI(0.5\theta_2 - 1.5u_2{}^\circ + 3\theta_3). & \text{(d)}
\end{aligned} \qquad (4.36)$$

As stated earlier, the 5 kN load is shared between members 1–2 and 3–4, therefore it will be necessary to add together (4.36(a)) and (4.36(c)), so that the resulting simultaneous equations will take the form of (4.37):

$$\begin{aligned}
5 &= EI(3u_2{}^\circ - 1.5\theta_2 - 1.5\theta_3) \\
1 &= EI(-1.5u_2{}^\circ + 3\theta_2 + 0.5\theta_3) \\
-2 &= EI(-1.5u_2{}^\circ + 0.5\theta_2 + 3\theta_3).
\end{aligned} \qquad (4.37)$$

Solving (4.37), the following are obtained:

$$\begin{aligned}
u_2{}^\circ &= 2.6667/EI = u_3{}^\circ \\
\theta_2 &= 1.6/EI \\
\theta_3 &= 0.4/EI
\end{aligned}$$

To determine the nodal moments

Member 1–2

$$s = 1; c = 0$$

$$v_2 = \lfloor -s \quad c \rfloor \begin{Bmatrix} u_2{}^{\circ} \\ v_2{}^{\circ} \end{Bmatrix} = -2.6667/EI$$

From the slope-deflection equations (4.8) and (4.10),

$$\begin{Bmatrix} M_1 \\ M_2 \end{Bmatrix} = EI \begin{bmatrix} -1.5 & 2 & 1.5 & 1 \\ -1.5 & 1 & 1.5 & 2 \end{bmatrix} \begin{Bmatrix} v_1 \\ \theta_1 \\ v_2 \\ \theta_2 \end{Bmatrix} + \begin{Bmatrix} \text{End fixing} \\ \text{moments} \end{Bmatrix}$$

$$= EI \begin{bmatrix} -1.5 & 2 & 1.5 & 1 \\ -1.5 & 1 & 1.5 & 2 \end{bmatrix} \frac{1}{EI} \begin{Bmatrix} 0 \\ 0 \\ -2.6667/EI \\ 1.6/EI \end{Bmatrix} + \begin{Bmatrix} -1 \\ 1 \end{Bmatrix}$$

$$= \begin{Bmatrix} -3.4 \\ 0.2 \end{Bmatrix} \text{kN m.}$$

Member 2–3

$$c = 1; s = 0.$$

$$v_2{}^{\circ} = \lfloor -s \quad c \rfloor \begin{Bmatrix} u_2{}^{\circ} \\ v_2{}^{\circ} \end{Bmatrix} = 0$$

$$v_3{}^{\circ} = \lfloor -s \quad c \rfloor \begin{Bmatrix} u_3{}^{\circ} \\ v_3{}^{\circ} \end{Bmatrix} = 0$$

From (4.8) and (4.10),

$$\begin{Bmatrix} M_2 \\ M_3 \end{Bmatrix} = EI \begin{bmatrix} -0.375 & 1 & 0.375 & 0.5 \\ -0.375 & 0.5 & 0.375 & 1 \end{bmatrix} \frac{1}{EI} \begin{Bmatrix} 0 \\ 1.6/EI \\ 0 \\ 0.4/EI \end{Bmatrix} + \begin{Bmatrix} -2 \\ 2 \end{Bmatrix}$$

$$= \begin{Bmatrix} -0.2 \\ 3.2 \end{Bmatrix} \text{kN m.}$$

Member 3–4

$$c = 0; s = -1.$$

$$v_3 = \lfloor -s \quad c \rfloor \begin{Bmatrix} u_3{}^{\circ} \\ v_3{}^{\circ} \end{Bmatrix} = 2.6667/EI.$$

$$\begin{Bmatrix} M_3 \\ M_4 \end{Bmatrix} = \begin{Bmatrix} -3.2 \\ 3.6 \end{Bmatrix} \text{kN m}$$

N.B. It should be noted that the nodal moments balance at nodes 2 and 3, as required.

4.4 STIFFNESS MATRIX FOR A TORQUE BAR

Prior to obtaining the elemental stiffness matrix for the general case of the one-dimensional member, the stiffness matrix for a shaft in torsion must be determined.

Consider a shaft of length '*l*' subjected to torques T_1 and T_2 at nodes 1 and 2 in a direction according to the right-hand screw rule (i.e., by pointing the right hand in a direction parallel to the arrows shown in Fig. 4.12, the directions of rotation are obtained by rotating the right hand in a clockwise direction).

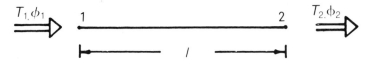

Fig. 4.12 – Shaft element.

From the relationship

$$T = \frac{GJ}{l}\phi$$

where

G = modulus of rigidity,
J = torsional constant,
ϕ = angle of twist,

we obtain

$$T_1 = \frac{GJ}{l}(\phi_1 - \phi_2),$$

and

$$T_2 = -T_1 = \frac{GJ}{l}(\phi_2 - \phi_1). \tag{4.38}$$

Rewriting (4.38) in matrix form, the following is obtained:

$$\begin{Bmatrix} T_1 \\ T_2 \end{Bmatrix} = \frac{GJ}{l} \begin{bmatrix} 1 & -1 \\ -1 & 1 \end{bmatrix} \begin{Bmatrix} \phi_1 \\ \phi_2 \end{Bmatrix}$$

or

$$\{P_i\} = [k]\{u_i\}$$

where

$$\{P_i\} = \begin{Bmatrix} T_1 \\ T_2 \end{Bmatrix}$$

$$\{u_i\} = \begin{Bmatrix} \phi_1 \\ \phi_2 \end{Bmatrix}$$

$$[k] = \frac{GJ}{l} \begin{bmatrix} 1 & -1 \\ -1 & 1 \end{bmatrix} = \text{elemental stiffness matrix for a torque bar.}$$

$$(4.39)$$

4.5 GENERAL CASE OF THE ONE-DIMENSIONAL MEMBER

The element for this case, in local coordinates, is shown in Fig. 4.13, where the double-tailed arrows represent torque and twist according to the right-hand screw rule.

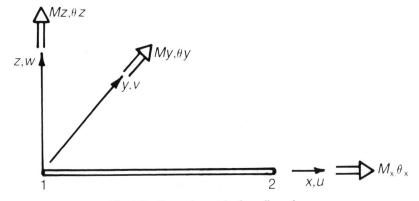

Fig. 4.13 – Beam element in three dimensions.

For this case, it is necessary to include bending, twisting, and axial stiffness, hence, from equations (3.15), (4.13), and (4.39), the elemental stiffness matrix for the general case of the one-dimensional member is given in

Table 4.1, where I_y and I_z are the second moments of area about the x–y and x–z planes.

The element is shown in global coordinates in Fig. 4.14, and by resolution, it can be shown that the relationship between local and global forces and moments is given by:

$$\{P_i\} = [\Xi]\{P_i^\circ\}$$

where

$$[\Xi] = \begin{bmatrix} \zeta & & & \\ \hline & \zeta & & \\ \hline & & \zeta & \\ \hline & & & \zeta \end{bmatrix}$$

and $[\zeta]$ is as equation (3.31).

$$\{P_i\}^{\mathrm{T}} = \lfloor X_1\ Y_1\ Z_1\ M_{x1}(=T_1)\ M_{y1}\ M_{z1}\ X_2\ Y_2\ Z_2\ T_2\ M_{y2}\ M_{z2}\rfloor$$
$$\{P_i^\circ\}^{\mathrm{T}} = \lfloor X_1^\circ\ Y_1^\circ\ Z_1^\circ\ M_{x1}^\circ\ M_{y1}^\circ\ M_{z1}^\circ\ X_2^\circ\ Y_2^\circ\ Z_2^\circ\ M_{x2}^\circ\ M_{y2}^\circ\ M_{z2}^\circ\rfloor.$$

Similarly,

$$\{u_i\} = [\Xi]\{u_i^\circ\}$$
$$\{u_i\}^{\mathrm{T}} = \lfloor u_1\ v_1\ w_1\ \theta_{x1}\ \theta_{y1}\ \theta_{z1}\ u_2\ v_2\ w_2\ \theta_{x2}\ \theta_{y2}\ \theta_{z2}\rfloor$$
$$\{u_i^\circ\}^{\mathrm{T}} = \lfloor u_1^\circ\ v_1^\circ\ w_1^\circ\ \theta_{x1}^\circ\ \theta_{y1}^\circ\ \theta_{z1}^\circ\ u_2^\circ\ v_2^\circ\ w_2^\circ\ \theta_{x2}^\circ\ \theta_{y2}^\circ\ \theta_{z2}^\circ\rfloor$$

and

$$[k^\circ] = [\Xi]^{\mathrm{T}}[k][\Xi]$$

= elemental stiffness matrix in global coordinates – this is normally determined numerically for each element, by a digital computer.

The directional cosines C_{x,x°, C_{x,y° and C_{x,z°, are calculated according to equations (3.33), whilst the directional cosines C_{y,x°, C_{y,y° and C_{y,z° are calculated by constructing a triangular plane through the centroid of the section in the x–y principal plane of bending, as shown in Figs 4.15 and 4.16. These directional cosines have been derived by Argyris [28] and Petyt [29], and will not be rederived here.

The expressions, however, for calculating the directional cosines will be given, and to assist their presentation, the following definitions will be made:

$$\{\psi_i^\circ\} = \begin{Bmatrix} x_i^\circ \\ y_i^\circ \\ z_i^\circ \end{Bmatrix}; \qquad \{\psi_j^\circ\} = \begin{Bmatrix} x_j^\circ \\ y_j^\circ \\ z_j^\circ \end{Bmatrix}$$

Table 4.1 Elemental stiffness matrix for the general case of the one-dimensional member

$$[k] = \begin{array}{c|cccccccccccc}
 & u_1 & v_1 & w_1 & \theta_{x1} & \theta_{y1} & \theta_{z1} & u_2 & v_2 & w_2 & \theta_{x2} & \theta_{y2} & \theta_{z2} \\
\hline
u_1 & \dfrac{AE}{l} & & & & & & & & & & & \\[2mm]
v_1 & 0 & \dfrac{12EI_z}{l^3} & & & & & & & & & \text{Symmetrical} & \\[2mm]
w_1 & 0 & 0 & \dfrac{12EI_y}{l^3} & & & & & & & & & \\[2mm]
\theta_{x1} & 0 & 0 & 0 & \dfrac{GJ}{l} & & & & & & & & \\[2mm]
\theta_{y1} & 0 & 0 & -\dfrac{6EI_y}{l^2} & 0 & \dfrac{4EI_y}{l} & & & & & & & \\[2mm]
\theta_{z1} & 0 & \dfrac{6EI_z}{l^2} & 0 & 0 & 0 & \dfrac{4EI_z}{l} & & & & & & \\[2mm]
u_2 & -\dfrac{AE}{l} & 0 & 0 & 0 & 0 & 0 & \dfrac{AE}{l} & & & & & \\[2mm]
v_2 & 0 & -\dfrac{12EI_z}{l^3} & 0 & 0 & 0 & -\dfrac{6EI_z}{l^2} & 0 & \dfrac{12EI_z}{l^3} & & & & \\[2mm]
w_2 & 0 & 0 & -\dfrac{12EI_y}{l^3} & 0 & \dfrac{6EI_y}{l^2} & 0 & 0 & 0 & \dfrac{12EI_y}{l^3} & & & \\[2mm]
\theta_{x2} & 0 & 0 & 0 & -\dfrac{GJ}{l} & 0 & 0 & 0 & 0 & 0 & \dfrac{GJ}{l} & & \\[2mm]
\theta_{y2} & 0 & 0 & -\dfrac{6EI_y}{l^2} & 0 & \dfrac{2EI_y}{l} & 0 & 0 & 0 & \dfrac{6EI_y}{l^2} & 0 & \dfrac{4EI_y}{l} & \\[2mm]
\theta_{z2} & 0 & \dfrac{6EI_z}{l^2} & 0 & 0 & 0 & \dfrac{2EI_z}{l} & 0 & -\dfrac{6EI_z}{l^2} & 0 & 0 & 0 & \dfrac{4EI_z}{l}
\end{array}$$

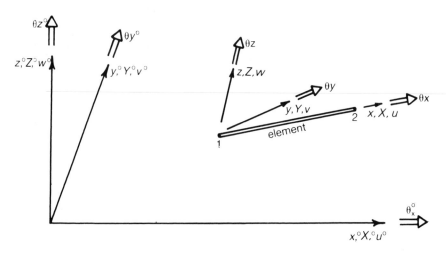

Fig. 4.14.

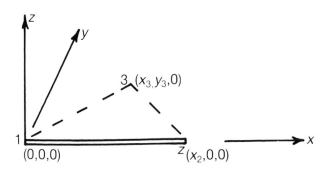

Fig. 4.15 – A principal plane of bending.

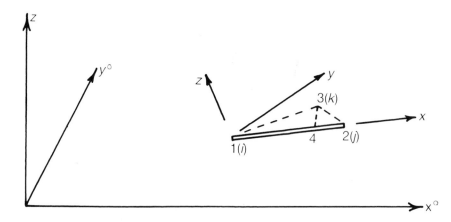

Fig. 4.16 – Triangular plane in local and global coordinates.

$$\{\psi_{ji}^\circ\} = \left\{ \begin{array}{c} x_j^\circ - x_i^\circ \\ y_j^\circ - y_i^\circ \\ z_j^\circ - z_i^\circ \end{array} \right\}; \qquad \{\psi_{ji}\} = \left\{ \begin{array}{c} x_j - x_i \\ y_j - y_i \\ z_j - z_i \end{array} \right\}$$

By constructing a line 3–4, as in Fig. 4.16, Petyt has shown that

$$\lfloor C_{y,x^\circ} \ C_{y,y^\circ} \ C_{y,z^\circ} \rfloor = \lfloor \psi_{34^\circ} \rfloor / l_{34^\circ}$$

where

$$(l_{34^\circ})^2 = \lfloor \psi_{34^\circ} \rfloor \{\psi_{34^\circ}\}$$
$$\{\psi_{34^\circ}\} = [\lfloor I \rfloor - \{\zeta_1\} \lfloor \zeta_1 \rfloor] \{\psi_{31^\circ}\}$$
$$\lfloor \zeta_1 \rfloor = \frac{1}{l_{21^\circ}} \{\psi_{21^\circ}\}^T = \lfloor C_{x,x^\circ} \ C_{x,y^\circ} \ C_{x,z^\circ} \rfloor$$
$$(l_{21^\circ})^2 = \{\psi_{21^\circ}\}^T \{\psi_{21^\circ}\}.$$

The x–z principal plane is perpendicular to the x–y principal plane and the o–z axis, and by using a cross-product of the two previously obtained vectors, Petyt has shown the directional cosines of the x–z plane to be given by:

$$\lfloor C_{z,x^\circ} \ C_{z,y^\circ}, \ C_{z,z^\circ} \rfloor = \frac{1}{\Delta} \lfloor \Delta_{y^\circ,z^\circ} \ \Delta_{z^\circ,x^\circ} \ \Delta_{x^\circ,y^\circ} \rfloor$$

where

Δ = area of triangle 123
$$= \sqrt{\lfloor \Delta_{y^\circ,z^\circ}^2 + \Delta_{z^\circ,x^\circ}^2 + \Delta_{x^\circ,y^\circ}^2 \rfloor}$$

Δ_{x°,y° = projected area of Δ_{123} on x°–y° plane

$$= \tfrac{1}{2} \begin{vmatrix} x_1^\circ & y_1^\circ & 1 \\ x_2^\circ & y_2^\circ & 1 \\ x_3^\circ & y_3^\circ & 1 \end{vmatrix}$$

Δ_{y°,z° = projected area of Δ_{123} on to plane y°–z°

$$= \tfrac{1}{2} \begin{vmatrix} y_1^\circ & z_1^\circ & 1 \\ y_2^\circ & z_2^\circ & 1 \\ y_3^\circ & z_3^\circ & 1 \end{vmatrix}$$

Δ_{z°,x° = projected area of Δ_{123} on to plane x°–z°

$$= \tfrac{1}{2} \begin{vmatrix} z_1^\circ & x_1^\circ & 1 \\ z_2^\circ & x_2^\circ & 1 \\ z_3^\circ & x_3^\circ & 1 \end{vmatrix}.$$

It is not, of course, very practical to demonstrate the application of the method to a space frame with a manual calculation, so the computer program of reference (15) will be used instead.

4.5.1 Example 4.6

Using the computer program '3D-RIGFRAME' of reference [15] or that of reference [27], determine the nodal displacements, forces, moments, etc., of the rigid-jointed space frame of Fig. 4.17. It may be assumed that the structure is constructed from 10 m lengths of piping, of external diameter 0.3 m and 2 cm wall thickness. $E = 2 \times 10^8$ kN/m², and $G = 7.7 \times 10^7$ kN/m².

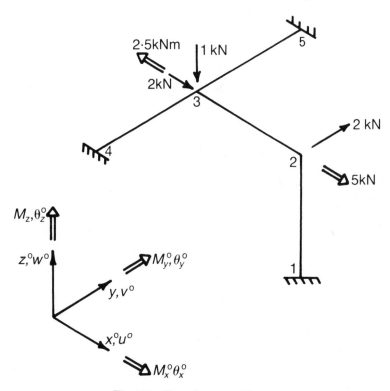

Fig. 4.17 – Three-dimensional frame.

Results
From '3D-RIGFRAME', the nodal displacements in global coordinates for the free nodes are as follows:

Node	metres			radians		
	u^o	v^o	w^o	θ_x^o	θ_y^o	θ_z^o
2	1.7719E-3	5.2667E-3	-8.3167E-7	-3.8127E-4	1.0197E-4	6.0021E-4
3	1.7734E-3	8.5621E-7	-8.5047E-4	-1.1420E-4	-1.3120E-4	1.6326E-4

The nodal forces, moments, etc., are as follows:

Member	Node	Axial Force (kN)	Torque (kN m)	M_{x-y} (kN m)	M_{x-z} (kN m)
1–2–3	1	− 0.293	1.498	8.309	− 2.978
1–2–3	2	− 0.293	1.498	− 5.666	2.271
3–2–1	3	− 0.525	− 0.666	− 4.527	− 0.655
3–2–1	2	− 0.525	− 0.666	1.498	2.271
4–3–2	4	0.301	− 0.327	− 0.977	4.819
4–3–2	3	0.301	− 0.327	0.185	− 5.951
5–3–2	5	− 0.301	0.327	2.560	2.556
5–3–2	3	− 0.301	0.327	− 3.352	− 1.424

The above analysis was carried out on an Apple II microcomputer, using the '3D-RIGFRAME' program of reference [15], and a plot of the deflected form of the framework is shown in Fig. 4.18.

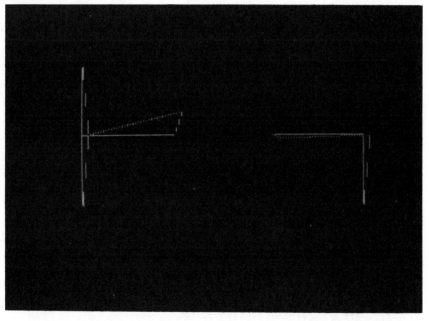

Fig. 4.18 – Deflected form of space frame (Plan on left, and front elevation on right of screen)

4.6   COMPUTER PROGRAMS

The computer programs "BEAMS", 'PLANEFRAME', and '3D-RIGFRAME', which are capable of analysing the structures discussed in this chapter, have been published in reference [15].

EXAMPLES FOR PRACTICE

1. Using the matrix displacement method, determine the nodal displacements and moments for the beams of 1(a) to 1(d).

(a)

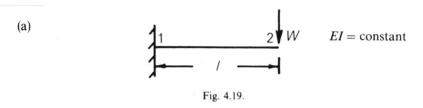

Fig. 4.19.

Answer

$v_2 = -Wl^3/(3EI)$

$M_1 = -Wl$

$M_2 = 0$

(b)

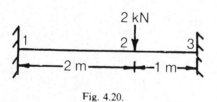

Fig. 4.20.

$E = 2E8 \text{ kN/m}^2 \qquad I = 0.01 \text{ m}^4$

Answer

$v_2 = -9.88E\text{-}8 \text{ m}$

$\theta_2 = -7.41E\text{-}8 \text{ rads}$

$M_1 = -0.444 \text{ kN m}$

$M_2 = \pm 0.59 \text{ kN m}$

$M_3 = 0.899 \text{ kN m}.$

(c)

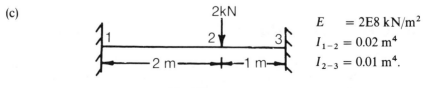

Fig. 4.21.

$E \quad = 2E8 \text{ kN/m}^2$

$I_{1-2} = 0.02 \text{ m}^4$

$I_{2-3} = 0.01 \text{ m}^4.$

Answer

$v_2 = -7.21\text{E-8 m}$

$\theta_2 = -2.7\text{E-8 rads}$

$M_1 = -0.541 \text{ kN m}$

$M_2 = \pm 0.649 \text{ kN m}$

$M_3 = 0.757 \text{ kN m}.$

(d)

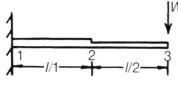

$E = \text{constant}$

$I_{1-2} = 2I$

$I_{2-3} = I$

Fig. 4.22.

Answer

$v_2 = -0.052 \ W l^3/EI$

$\theta_2 = 0.188 \ W l^2/EI$

$v_3 = -0.188 \ W l^2/EI$

$\theta_3 = 0.312 \ W l^2/EI$

$M_1 = -W l$

$M_2 = -W l/2$

$M_3 = 0.$

2. Using the matrix displacement method, determine the nodal displacements and moments for the beams of 2(a) to 2(c).

(a) $EI = \text{constant}$

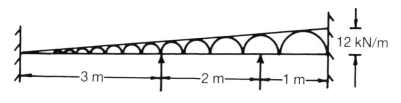

Fig. 4.23.

The end fixing moments for the hydrostatically loaded beam shown in Fig. 4.24 are as follows:

$M_A^F = -l^2(w_1/20 + w_2/30)$

$M_B^F = l^2(w_1/30 + w_2/20).$

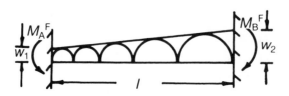

Fig. 4.24.

Answer

$\theta_2 = 0.047/EI$

$\theta_3 = -0.3235/EI$

$M_1 = -1.769 \text{ kN m}$

$M_2 = \pm 2.76 \text{ kN m}$

$M_3 = \pm 2.2 \text{ kN m}$

$M_4 = 0.28 \text{ kN m}.$

(Portsmouth Polytechnic, 1978)

(b)

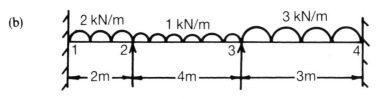

Fig. 4.25.

$EI = \text{constant}$

Answer

$\theta_2 = 0.1626/EI$

$\theta_3 = 0.3580/EI$

$M_1 = -0.504 \text{ kN/m}$

$M_2 = \pm 0.992 \text{ kN m}$

$M_3 = \pm 1.773 \text{ kN m}$

$M_4 = 2.489 \text{ kN m}.$

(Portsmouth Polytechnic, 1977)

(c)

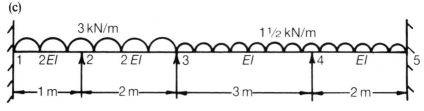

Fig. 4.26.

Answer

$$\lfloor \theta_2 \quad \theta_3 \quad \theta_4 \rfloor = \frac{1}{EI} \lfloor 0.0582 \quad 0.02569 \quad -0.1926 \rfloor$$

$M_1 = -0.0172 \text{ kN m}$

$M_2 = \pm 0.7156 \text{ kN m}$

$M_3 = \pm 1.2192 \text{ kN m}$

$M_4 = \pm 0.8853 \text{ kN m}$

$M_5 = 0.3074 \text{ kN m}.$

(Portsmouth Polytechnic, 1979)

3. Determine the nodal displacements and moments for the rigid-jointed planes frames of 3(a) and 3(b). In both cases, it may be assumed that the axial stiffness is very large compared with the flexural stiffness.

(a)

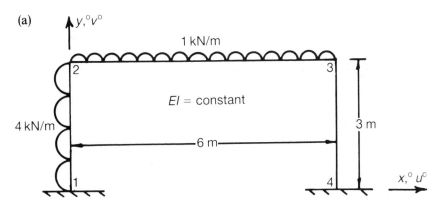

Fig. 4.27.

Assumptions: $v_2^\circ = v_3^\circ = 0$ and $u_2^\circ = u_3^\circ$.

Answer

$u_2^\circ = 24.3/EI$

$\theta_2 = 7.2/EI$

$\theta_3 = 5.46/EI$

$M_1 = -14.4 \text{ kN m}$

$M_2 = \pm 3.6 \text{ kN m}$

$M_3 = \pm 9 \text{ kN m}$

$M_4 = 12.6 \text{ kN m}.$

(Portsmouth Polytechnic, 1980)

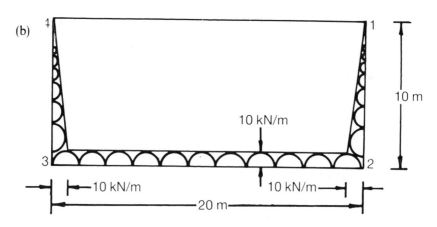

Fig. 4.28.

Assumptions: $u_1^\circ = u_2^\circ = u_3^\circ = u_4^\circ = 0$ and $v_1^\circ = v_2^\circ = v_3^\circ = v_4^\circ = 0$
For element 1–2, $EI = 10\,\text{kN m}^2$
For elements 2–3 and 1–4, $EI = 20\,\text{kN m}^2$
For element 3–4, $EI = 100\,\text{kN m}^2$.
The formulae for end fixing moments are as for 2(a).

Answer

$\theta_1 = -\theta_2 = 3.65$
$\theta_3 = -\theta_4 = 16.55$
$M_1 = 3.65\,\text{kN m}$
$M_2 = \pm 3.65\,\text{kN m}$
$M_3 = \pm 167.8\,\text{kN m}$
$M_4 = 167.8\,\text{kN m}$.

(Portsmouth Polytechnic, 1982)

Chapter 5

Finite element analysis

Although the matrix displacement method is part and parcel of the finite element method, the latter description is usually associated with the theory presented in the present chapter, which was first introduced by Turner *et al.* [5].

For plates and shells with complex shapes and complex load systems, where analysis by traditional methods is very difficult, or virtually impossible, the method described in the present chapter has been found to be particularly useful. For structures of this type, it is usual practice to represent their shapes with a large number of smaller shapes, known as finite elements.

Finite elements take many and varied forms, depending on the shape they are supposed to represent. For example, to represent flat plates, the choice of finite elements will usually be of triangular or of quadrilateral shape (Fig. 5.1), whilst for solids, the finite elements will usually appear in the form of tetrahedrons or cubes.

In Fig. 5.1, the finite elements are described by corner nodes, but more sophisticated elements with curved boundaries and mid-side nodes are gaining popularity, as shown in Fig. 5.2, as they usually lend themselves to a more satisfactory solution.

As the displacement method is normally used in finite element analysis, it is evident that one of the main problems to be overcome will be the determination of the element stiffness matrices. This is because although elements are usually of simple shape in comparison with the overall structure, their shapes are quite complicated when compared with the usual skeletal elements.

Because of this, it has been found desirable to derive these stiffness

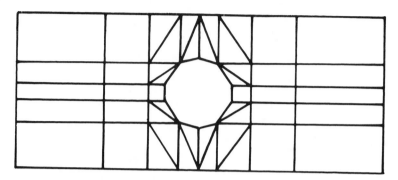

Fig. 5.1 – Mesh for a flat plate with a hole.

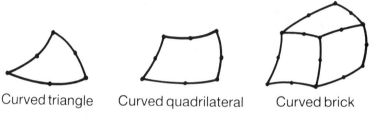

Curved triangle   Curved quadrilateral   Curved brick

Fig. 5.2 – Elements with mid-side nodes.

matrices by various energy methods [26], and two of these, namely the virtual work method and the method of minimum potential, are used to derive stiffness matrices for one- and two-dimensional elements. The chapter also deals with thermal stresses and distributed loads, and gives a method for reducing the number of simultaneous equations through reduction of the system stiffness matrices.

Dynamic and stability problems, which can also be tackled by the finite element method, are considered in Chapters 7, 8, and 9.

5.1 DERIVATION OF ELEMENTAL STIFFNESS MATRICES

5.1.1 Method 1 – virtual work approach

During the process of describing the method, it will also be applied to the *plane rod element* of Fig. 5.3. It is hoped that this will assist with obtaining a better understanding of the method.

Let $\{u_{(x, y)}\}$ = a vector representing displacement functions of the element. For the plane rod, it will be convenient to assume the following displacement function:

$$u = \alpha_1 + \alpha_2 x = \{u_{(x, y)}\}, \tag{5.1}$$

where α_1 and α_2 are constants.

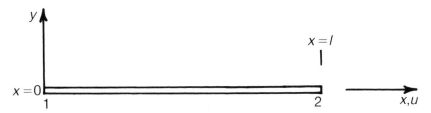

Fig. 5.3 – Plane rod element.

As there are two boundary values for displacement, namely u_1 (at $x = 0$) and u_2 (at $x = l$), the assumption for the displacement function 'u' can only include the two unknown constants α_1 and α_2. It is necessary to assume the same number of constants 'α' as there are nodal degrees of freedom for the element, so that the correct number of simultaneous equations can be obtained to determine the 'α's' in terms of the nodal displacements, $\{u_i\}$.

In matrix form, (5.1) appears as

$$\{u_{(x, y)}\} = [1 \quad x]\begin{Bmatrix} \alpha_1 \\ \alpha_2 \end{Bmatrix}$$

$$= [M_{(x, y)}]\{\alpha_i\} \tag{5.2}$$

where

$[M_{(x, y)}] =$ a matrix containing x, y, z, etc., and their various powers, depending on the assumed polynomials for $\{u_{(x, y)}\}$. In many cases, $[M_{(x, y)}]$ will be rectangular, because these elements are dependent on a number of displacement functions.

$\{\alpha_i\} =$ a vector of unknown constants.

As $\{\alpha_i\}$ are unknown, it is necessary to obtain these in terms of the nodal displacements, as follows:

$$@ \ x = 0, \quad u = u_1$$

and

$$@ \ x = l, \quad u = u_2. \tag{5.3}$$

Substituting (5.3) into (5.1), the following is obtained:

$$u_1 = \alpha_1$$

$$u_2 = \alpha_1 + \alpha_2 l. \tag{5.4}$$

Rewriting (5.4) into matrix form:

$$\begin{Bmatrix} u_1 \\ u_2 \end{Bmatrix} = \begin{bmatrix} 1 & 0 \\ 1 & l \end{bmatrix}\begin{Bmatrix} \alpha_1 \\ \alpha_2 \end{Bmatrix}$$

or

$$\{u_i\} = [C]\{\alpha_i\} \tag{5.5}$$

where

$\{u_i\}$ = a vector of nodal displacements
$[C]$ = a matrix of constants ($[C]$ is always square, and its order is the same as the degrees of freedom of the element and the number of unknown polynomial coefficients 'α')

In this case,

$$[C] = \begin{bmatrix} 1 & 0 \\ 1 & l \end{bmatrix}. \tag{5.6}$$

From (5.5) the unknown $\{\alpha_i\}$ can be obtained as follows:

$$\{\alpha_i\} = [C^{-1}]\{u_i\}. \tag{5.7}$$

In this case,

$$[C^{-1}] = \begin{bmatrix} 1 & 0 \\ -1/l & 1/l \end{bmatrix}. \tag{5.8}$$

As the virtual work equation of Chapter 2 involves strain, it will be necessary to obtain an expression for strain in terms of the above matrices.

Now, for a plane rod,

$$\varepsilon = \frac{du}{dx} = \frac{d\{u_{(x,\,y)}\}}{dx} = \alpha_2$$

which in matrix form appears as

$$\{\varepsilon\} = [0 \quad 1]\begin{Bmatrix} \alpha_1 \\ \alpha_2 \end{Bmatrix}$$

$$= [B_{(x,\,y)}]\{\alpha_i\} \tag{5.9}$$

where

$[B_{(x,\,y)}]$ = a matrix relating a vector of strains and the unknown vector $\{\alpha_i\}$.

In this case,

$$[B_{(x,\,y)}] = [0 \quad 1]. \tag{5.10}$$

Substituting (5.7) into (5.9),

$$\{\varepsilon\} = [B_{(x,\,y)}][C]^{-1}\{u_i\}. \tag{5.11}$$

Now for uniaxial stress,

$$\sigma = E\varepsilon$$

or in matrix form, for two- and three-dimensional stress systems,

$$\{\sigma\} = [D]\{\varepsilon\} \tag{5.12}$$

where

$$\{\sigma\} = \text{a vector of internal stresses}$$

and

$$[D] = \text{a matrix of material constants.}$$

In this case,

$$[D] = E = \text{Young's modulus.} \tag{5.13}$$

Substituting (5.11) into (5.12),

$$\{\sigma\} = [D][B_{(x, y)}][C]^{-1}\{u_i\}.$$

Now, from Chapter 2,

$$\text{internal virtual work} = \int_{\text{vol}} \{\bar{\varepsilon}\}^{\text{T}}\{\sigma\}\,d(\text{vol})$$

where

$$\{\bar{\varepsilon}\} = \text{a vector of virtual strains.}$$

On substitution:
Internal virtual work

$$= \int_{\text{vol}} \{\bar{u}_i\}^{\text{T}}[C^{-1}]^{\text{T}}[B_{(x, y)}]^{\text{T}}[D][B_{(x, y)}][C^{-1}]\{u_i\}\,d(\text{vol})$$

$$= \{\bar{u}_i\}^{\text{T}}[C^{-1}]^{\text{T}}\int_{\text{vol}}[B_{(x, y)}]^{\text{T}}[D][B_{(x, y)}][C^{-1}]\{u_i\}\,d(\text{vol}) \tag{5.14}$$

where

$$\{\bar{u}_i\} = \text{a vector of virtual nodal displacements}$$

$$\text{External virtual work done on the element} = \{\bar{u}_i\}^{\text{T}}\{p_i\} \tag{5.15}$$

Substituting (5.14) and (5.15) into (2.18), the following relationship is obtained:

$$\{p_i\} = [C^{-1}]^{\text{T}}\int_{\text{vol}}[B_{(x, y)}]^{\text{T}}[D][B_{(x, y)}]\,d(\text{vol})[C^{-1}]\{u_i\}$$

$$= [k]\{u_i\}.$$

Hence, the elemental stiffness matrix can be expressed in the form

$$[k] = [C^{-1}]^{\text{T}}[\tilde{k}][C^{-1}] \tag{5.16}$$

where

$$[\tilde{k}] = \int_{vol} [B_{(x,\,y)}]^T [D] [B_{(x,\,y)}] \, d\,(vol)$$ (5,17)

For the rod element, $[\tilde{k}]$ is obtained by substituting (5.10) and (5.13) into (5.17), as follows:

$$[\tilde{k}] = \int_0^l A \begin{bmatrix} 0 \\ 1 \end{bmatrix} [E] [0 \quad 1] \, dx$$

$$= Al \begin{bmatrix} 0 & 0 \\ 0 & E \end{bmatrix}$$ (5.18)

where
   A = cross-sectional area of rod.

Substituting (5.8) and (5.18) into (5.16), the following is obtained for the elemental stiffness matrix for a rod:

$$[k] = \begin{bmatrix} 1 & -1/l \\ 0 & 1/l \end{bmatrix} Al \begin{bmatrix} 0 & 0 \\ 0 & E \end{bmatrix} \begin{bmatrix} 1 & 0 \\ -1/l & 1/l \end{bmatrix}$$

$$= \frac{AE}{l} \begin{bmatrix} 1 & -1 \\ -1 & 1 \end{bmatrix} \text{ (as required).}$$

From the expression for internal virtual work, it can be seen that

$$\int_{vol} \{\bar{\varepsilon}\}^T \{\sigma\} \, d\,(vol) = \{\bar{u}_i\}^T [k] \{u_i\} \text{ (see Chapter 2).}$$

Similarly,

$$U_e = \frac{1}{2} \int_{vol} \{\varepsilon\}^T \{\sigma\} \, d\,(vol)$$

$$= \frac{1}{2} \{u_i\}^T [k] \{u_i\}$$ (5.19)

$$= \text{strain energy (see Chapter 2).}$$

5.1.2   Method 2 – Minimum potential approach

For many problems this is often simpler than the virtual work approach. From section 2.7, the total potential

$$\pi_p = U_e + W.$$

From (5.19),

$$U_e = \frac{1}{2}\{u_i\}^T [k] \{u_i\}.$$

Hence,

$$\pi_p = \frac{1}{2}\{u_i\}^T [k] \{u_i\} - \{u\}^T \{p_i\}.$$

From the method of minimum potential,

$$\frac{\partial \pi_p}{\partial \{u_i\}} = \{0\}$$

or

$$\{0\} = [k] \{u_i\} - \{p_i\},$$

therefore

$$\frac{\partial U_e}{\partial \{u_i\}} = [k] \{u_i\}. \tag{5.20}$$

Thus, the stiffness matrix can be obtained by differentiating the strain energy with respect to the nodal displacements.

To determine $[k]$ for a rod from equation (5.20)
Assume as before,

$$u_{(x, y)} = \alpha_1 + \alpha_2 x.$$

At $x = 0$,

$$u_{(x, y)} = u_1 = \alpha_1$$

At $x = l$,

$$u_{(x, y)} = u_2 = \alpha_1 + \alpha_2 l,$$

therefore

$$\alpha_2 = \frac{u_2 - u_1}{l},$$

that is,

$$u_{(x, y)} = u_1 + (u_2 - u_1)\frac{x}{l}. \tag{5.21}$$

The strain energy stored in the bar is given by,

$$U_e = \frac{E}{2} \int \frac{du_{(x,\,y)}^2}{dx} \, d\,(vol)$$

$$= \frac{AE}{2} \int_0^l \frac{du_{(x,\,y)}^2}{dx} \, dx$$

$$= \frac{AE}{2} \int_0^l \frac{(u_2 - u_1)^2}{l^2} dx$$

$$= \frac{AE}{2l} (u_2^2 + u_1^2 - 2u_1 u_2)$$

$$\frac{\partial U_e}{\partial u_1} = \frac{AE}{2l} (2u_1 - 2u_2) = \frac{AE}{l} \lfloor 1 \quad -1 \rfloor \begin{Bmatrix} u_1 \\ u_2 \end{Bmatrix}$$

$$\frac{\partial U_e}{\partial u_2} = \frac{AE}{l} (2u_2 - 2u_1) = \frac{AE}{l} \lfloor -1 \quad 1 \rfloor \begin{Bmatrix} u_1 \\ u_2 \end{Bmatrix}.$$

(5.22)

From (5.22) it can readily be seen that the elemental stiffness matrix is given by:

$$[k] = \frac{AE}{l} \begin{bmatrix} 1 & -1 \\ -1 & 1 \end{bmatrix}.$$

A *more sophisticated approach* with the method of minimum potential will now be given, for determining $[k]$.

Equation (5.21) can be rewritten in the form:

$$u_{(x,\,y)} = u_1 (1 - x/l) + u_2 (x/l)$$

or in matrix form,

$$\{u_{(x,\,y)}\} = [(1 - x/l) \quad (x/l)] \begin{Bmatrix} u_1 \\ u_2 \end{Bmatrix}$$

$$= [N] \{u_i\}$$

(5.23)

where $[N]$ is known as a matrix of shape functions.

On comparison of (5.23) with (5.2), it can be seen that

$$[M_{(x,\,y)}] \{\alpha_i\} = [N] \{u_i\}$$

or

$$[M_{(x,\,y)}] [C^{-1}] \{u_i\} = [N] \{u_i\}$$

or

$$[N] = [M_{(x,\,y)}] [C^{-1}]$$

(5.24)

Similarly, the expression

$$\{\varepsilon\} = [B_{(x, y)}]\{\alpha_i\} = [B_{(x, y)}][C^{-1}]\{u_i\}$$

can be represented by

$$\{\varepsilon\} = [B]\{u_i\} \qquad (5.25)$$

where

$$[B] \text{ is a function of } [N] = [B_{(x, y)}][C^{-1}]$$

and

$$[B_{(x, y)}] \text{ is a function of } [M_{(x, y)}].$$

Now,

$$U_e = \int_{vol} \frac{\{\sigma\}^T\{\varepsilon\}}{2} \, d(vol)$$

and

$$\{\sigma\}^T = \{\varepsilon\}^T[D],$$

so that

$$U_e = \frac{1}{2} \int_{vol} \{\varepsilon\}^T[D]\{\varepsilon\} \, d(vol)$$

$$= \frac{1}{2} \int_{vol} \{u_i\}^T[B]^T[D][B]\{u_i\} \, d(vol)$$

$$= \frac{1}{2} \{u_i\}^T[k]\{u_i\},$$

therefore

$$[k] = \int_{vol} [B]^T[D][B] \, d(vol). \qquad (5.26)$$

This method is particularly suitable if the relationship between the assumed displacement functions and the nodal displacements is known.

To Determine [k] for a rod from equation (5.26)
Equation (5.26) will now be used to obtain the elemental stiffness matrix for a rod. From (5.23),

$$\{u_{(x, y)}\} = [N]\{u_i\}$$

$$= [(1 - x/l) \quad x/l]\begin{Bmatrix} u_1 \\ u_2 \end{Bmatrix}$$

and

$$\{\varepsilon\} = \frac{d\{u_{(x, y)}\}}{dx}$$

$$= [-1/l \quad 1/l] \begin{Bmatrix} u_1 \\ u_2 \end{Bmatrix}$$

$$= [B]\{u_i\}$$

where

$$[B] = [-1/l \quad 1/l]. \tag{5.27}$$

Substituting (5.13) and (5.27) into (5.26), the following is obtained for the elemental stiffness matrix for a rod:

$$[k] = \int_0^l \begin{bmatrix} -1/l \\ 1/l \end{bmatrix} E[-1/l \quad 1/l] \, A dx = AEl \begin{bmatrix} 1/l^2 & -1/l^2 \\ -1/l^2 & 1/l^2 \end{bmatrix}$$

$$= \frac{AE}{l} \begin{bmatrix} 1 & -1 \\ -1 & 1 \end{bmatrix} - \text{as required.}$$

The result can be seen to be the same as before, but considerably simpler than the other derivations.

5.2 PLANE STRESS AND PLANE STRAIN PROBLEMS

Plane stress [30, 31] is a two-dimensional system of stress, where the stresses act in the plane of the plate. The resulting strains, due to the two-dimensional system of stress, will be three-dimensional, where the out-of-plane strain will be caused by the Poisson effect of the in-plane stresses.

Plane strain is a two-dimensional system of strain and a three-dimensional system of stress. In general, the two main stresses will be in-plane, and the third stress, namely the out-of-plane stress, will be related to the in-plane stresses and Poisson's ratio. The out-of-plane stress acts in such a manner that the out-of-plane strain is zero. Plane strain is of much importance in the rolling of metals and in the testing of materials to resist fatigue failure.

Apart from analysing plates with holes, such as shown in Fig. 5.1, in-plane plate elements have been successfully used to determine the strength of dams [11] and supertankers [32], as diagrammatically illustrated by Figs 5.4 and 5.5.

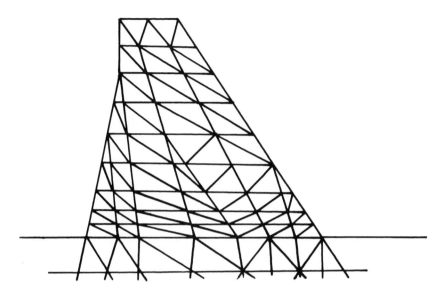

Fig. 5.4 – Cross-section of an earth dam.

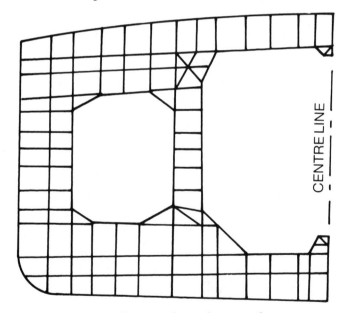

Fig. 5.5 – Transverse frame of a supertanker.

5.2.1 In-plane triangular element

First, the *virtual work principle* is used to derive the stiffness matrix for a flat triangular plate under plane stress, and later, this stiffness matrix is derived by the method of minimum potential.

Consider the in-plane triangular plate element of Fig. 5.6, which was first analysed by Turner *et al.* [5], by this method. It is a constant stress element and has three nodal points at its corners.

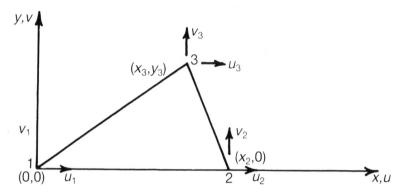

Fig. 5.6 – Flat element under in-plane forces.

The following displacement configurations will be assumed, and it should be noted that the six constants α_1 to α_6 correspond to the six nodal degrees of freedom. (It is necessary to assume the same number of constants as the number of nodal degrees of freedom per element, so that the number of simultaneous equations to be solved will be the same as the number of unknown α's, i.e. $[C]$ must be square).

$$u = \alpha_1 + \alpha_2 x + \alpha_3 y$$
$$v = \alpha_4 + \alpha_5 x + \alpha_6 y \tag{5.28}$$

Putting (5.28) into matrix form,

$$\left\{ \begin{matrix} u \\ v \end{matrix} \right\} = \{u_{(x,\,y)}\} = \begin{bmatrix} 1 & x & y & 0 & 0 & 0 \\ 0 & 0 & 0 & 1 & x & y \end{bmatrix} \left\{ \begin{matrix} \alpha_1 \\ \alpha_2 \\ \alpha_3 \\ \alpha_4 \\ \alpha_5 \\ \alpha_6 \end{matrix} \right\} \tag{5.29}$$

Comparing (5.29) with (5.2), it can be seen that

$$[M_{(x,\,y)}] = \begin{bmatrix} 1 & x & y & 0 & 0 & 0 \\ 0 & 0 & 0 & 1 & x & y \end{bmatrix}. \tag{5.30}$$

To obtain $[C]$, it will be necessary to consider the nodal displacements, as these are effectively the boundary conditions for the element.

@ $x = y = 0$, $u = u_1 = \alpha_1$

and

$$v = v_1 = \alpha_4$$

@ $x = x_2$ and $y = 0$, $u = u_2 = \alpha_1 + \alpha_2 \, x_2$

and

$$v = v_2 = \alpha_4 + \alpha_5 \, x_2 \tag{5.31}$$

@ $x = x_3$ and $y = y_3$, $u = u_3 = \alpha_1 + \alpha_2 x_3 + \alpha_3 y_3$

and

$$v = v_3 = \alpha_4 + \alpha_5 x_3 + \alpha_6 y_3.$$

Rewriting (5.31) in matrix form,

$$
\begin{Bmatrix} u_1 \\ v_1 \\ u_2 \\ v_2 \\ u_3 \\ v_3 \end{Bmatrix}
=
\begin{bmatrix}
1 & 0 & 0 & 0 & 0 & 0 \\
0 & 0 & 0 & 1 & 0 & 0 \\
1 & x_2 & 0 & 0 & 0 & 0 \\
0 & 0 & 0 & 1 & x_2 & 0 \\
1 & x_3 & y_3 & 0 & 0 & 0 \\
0 & 0 & 0 & 1 & x_3 & y_3
\end{bmatrix}
\begin{Bmatrix} \alpha_1 \\ \alpha_2 \\ \alpha_3 \\ \alpha_4 \\ \alpha_5 \\ \alpha_6 \end{Bmatrix}.
\tag{5.32}
$$

Comparing (5.32) with (5.5), it can be seen that

$$
[C] =
\begin{bmatrix}
1 & 0 & 0 & 0 & 0 & 0 \\
0 & 0 & 0 & 1 & 0 & 0 \\
1 & x_2 & 0 & 0 & 0 & 0 \\
0 & 0 & 0 & 1 & x_2 & 0 \\
1 & x_3 & y_3 & 0 & 0 & 0 \\
0 & 0 & 0 & 1 & x_3 & y_3
\end{bmatrix}.
\tag{5.33}
$$

At first glance of (5.33), it may appear to be quite a formidable task to invert $[C]$, but if it is rewritten in the following form, the inversion becomes a relatively simple problem.

$$
[C_A] =
\left[
\begin{array}{c|c}
A & 0_3 \\
\hline
0_3 & A
\end{array}
\right]
$$

and

$$
[C_A]^{-1} =
\left[
\begin{array}{c|c}
A^{-1} & 0_3 \\
\hline
0_3 & A^{-1}
\end{array}
\right]
$$

where

$$[A] = \begin{bmatrix} 1 & 0 & 0 \\ 1 & x_2 & 0 \\ 1 & x_3 & y_3 \end{bmatrix}.$$

Once the inverse of $[A]$ is obtained, $[C]^{-1}$ can be obtained by rearranging $[C_A]^{-1}$. It should be noted that $[A]$ is a lower triangular matrix, and its inverse is another lower triangular matrix.

Now, from equation (1.27),

$$[A]^c = \begin{bmatrix} x_2 y_3 & -y_3 & x_{32} \\ 0 & y_3 & -x_3 \\ 0 & 0 & x_2 \end{bmatrix},$$

and from equation (1.28),

$$[A^{-1}] = \frac{[A]^{cT}}{|A|}$$

$$= \frac{\begin{bmatrix} x_2 y_3 & 0 & 0 \\ -y_3 & y_3 & 0 \\ x_{32} & -x_3 & x_2 \end{bmatrix}}{x_2 y_3}$$

where

$$x_{32} = x_3 - x_2$$

therefore

$$[C_A]^{-1} = \frac{1}{x_2 y_3} \begin{array}{c} \begin{matrix} u_1 & \quad u_2 & \quad u_3 & \quad v_1 & \quad v_2 & \quad v_3 \end{matrix} \\ \left[\begin{array}{ccc|ccc} x_2 y_3 & 0 & 0 & 0 & 0 & 0 \\ -y_3 & y_3 & 0 & 0 & 0 & 0 \\ x_{32} & -x_3 & x_2 & 0 & 0 & 0 \\ \hline 0 & 0 & 0 & x_2 y_3 & 0 & 0 \\ 0 & 0 & 0 & -y_3 & y_3 & 0 \\ 0 & 0 & 0 & x_{32} & -x_3 & x_2 \end{array} \right] \begin{matrix} u_1 \\ u_2 \\ u_3 \\ v_1 \\ v_2 \\ v_3. \end{matrix} \end{array}$$

Hence, by rearranging $[C_A]^{-1}$,

$$[C^{-1}] = \frac{1}{x_2 y_3} \begin{matrix} u_1 & v_1 & u_2 & v_2 & u_3 & v_3 \\ \begin{bmatrix} x_2 y_3 & 0 & 0 & 0 & 0 & 0 \\ 0 & x_2 y_3 & 0 & 0 & 0 & 0 \\ -y_3 & 0 & y_3 & 0 & 0 & 0 \\ 0 & -y_3 & 0 & y_3 & 0 & 0 \\ x_{32} & 0 & -x_3 & 0 & x_2 & 0 \\ 0 & x_{32} & 0 & -x_3 & 0 & x_2 \end{bmatrix} & \begin{matrix} u_1 \\ v_1 \\ u_2 \\ v_2 \\ u_3 \\ v_3 \end{matrix} \end{matrix} \quad (5.34)$$

To obtain $[B_{(x,y)}]$, it will be necessary to consider the components of strain.

From [30], the three components of strain in a two-dimensional system are given by:

$$\varepsilon_x = \frac{\partial u}{\partial x} = \text{direct strain in the } x \text{ direction}$$

$$\varepsilon_y = \frac{\partial v}{\partial y} = \text{direct strain in the } y \text{ direction}$$

$$\gamma_{xy} = \frac{\partial u}{\partial y} + \frac{\partial v}{\partial x} = \text{shear strain in the } x\text{–}y \text{ plane.}$$

In matrix form, these components of strain are given by:

$$\begin{Bmatrix} \varepsilon_x \\ \varepsilon_y \\ \gamma_{xy} \end{Bmatrix} = \begin{bmatrix} \dfrac{\partial}{\partial x} & 0 \\ 0 & \dfrac{\partial}{\partial y} \\ \dfrac{\partial}{\partial y} & \dfrac{\partial}{\partial x} \end{bmatrix} \begin{Bmatrix} u \\ v \end{Bmatrix}. \quad (5.35)$$

Hence from (5.28), $\varepsilon_x = \alpha_2$, $\varepsilon_y = \alpha_6$ and $\gamma_{xy} = \alpha_3 + \alpha_5$, which, when substituted into (5.35), appear as

$$\begin{Bmatrix} \varepsilon_x \\ \varepsilon_y \\ \gamma_{xy} \end{Bmatrix} = \begin{bmatrix} 0 & 1 & 0 & 0 & 0 & 0 \\ 0 & 0 & 0 & 0 & 0 & 1 \\ 0 & 0 & 1 & 0 & 1 & 0 \end{bmatrix} \begin{Bmatrix} \alpha_1 \\ \alpha_2 \\ \alpha_3 \\ \alpha_4 \\ \alpha_5 \\ \alpha_6 \end{Bmatrix} \quad (5.36)$$

comparing (5.36) with (5.9),

$$[B_{(x,y)}] = \begin{bmatrix} 0 & 1 & 0 & 0 & 0 & 0 \\ 0 & 0 & 0 & 0 & 0 & 1 \\ 0 & 0 & 1 & 0 & 1 & 0 \end{bmatrix}. \quad (5.37)$$

To obtain [D]

It can be readily shown that the stress–strain relationships for *plane stress* are given by

$$\varepsilon_x = \frac{\sigma_x}{E} - \frac{v\sigma_y}{E}$$

$$\varepsilon_y = \frac{\sigma_y}{E} - \frac{v\sigma_x}{E}$$

$$\gamma_{xy} = \tau_{xy}/G = 2(1+v)\,\tau_{xy}/E,$$

which can be written in the form

$$\sigma_x = \frac{E}{(1-v^2)}\,(\varepsilon_x + v\,\varepsilon_y)$$

$$\sigma_y = \frac{E}{(1-v^2)}\,(\varepsilon_y + v\,\varepsilon_x)$$

$$\tau_{xy} = \frac{E}{2(1+v)}\,\gamma_{xy} = \frac{E}{(1-v^2)}\left(\frac{1-v}{2}\right)\gamma_{xy}.$$

In matrix form, these equations appear as

$$\left\{\begin{matrix} \sigma_x \\ \sigma_y \\ \tau_{xy} \end{matrix}\right\} = \frac{E}{(1-v^2)}\begin{bmatrix} 1 & v & 0 \\ v & 1 & 0 \\ 0 & 0 & (1-v)/2 \end{bmatrix}\left\{\begin{matrix} \varepsilon_x \\ \varepsilon_y \\ \gamma_{xy} \end{matrix}\right\}. \tag{5.38}$$

Comparing (5.38) with (5.12), it can be seen that

$$[D] = \frac{E}{(1-v^2)}\begin{bmatrix} 1 & v & 0 \\ v & 1 & 0 \\ 0 & 0 & (1-v)/2 \end{bmatrix}. \tag{5.39}$$

For *plane strain*, however, (5.39) cannot be used, because the stress–strain relationships are different. For this case, the strain in the direction perpendicular to the x–y plane of the plate, namely ε_z, is zero, and to achieve this, it is necessary for a stress σ_z to act in this direction. The stress σ_z will be proportional to the stresses σ_x and σ_y, as follows:

$$\varepsilon_z = 0 = \frac{\sigma_z}{E} - \frac{v\sigma_x}{E} - \frac{v\sigma_y}{E},$$

therefore

$$\sigma_z = v\sigma_x + v\sigma_y.$$

The other stress–strain relationships are:

$$\varepsilon_x = \frac{\sigma_x}{E} - \frac{v\sigma_y}{E} - \frac{v\sigma_z}{E}$$

$$\varepsilon_y = \frac{\sigma_y}{E} - \frac{v\sigma_x}{E} - \frac{v\sigma_z}{E}$$

and

$$\gamma_{xy} = \tau_{xy}/G = 2\,(1+v)\,\tau_{xy}/E.$$

Using a similar process to that adopted for plane stress, it can be shown that the matrix of elastic constants is given by

$$[D] = \frac{E\,(1-v)}{(1+v)\,(1-2v)} \begin{bmatrix} 1 & v/(1-v) & 0 \\ v/(1-v) & 1 & 0 \\ 0 & 0 & (1-2v)/[2(1-v)] \end{bmatrix}. \tag{5.40}$$

Thus, in general,

$$[D] = E^1 \begin{bmatrix} 1 & \mu & 0 \\ \mu & 1 & 0 \\ 0 & 0 & \gamma \end{bmatrix} \tag{5.41}$$

where, for *plane stress,*

$$E^1 = E/(1-v^2)$$
$$\mu = v$$
$$\gamma = (1-v)/2$$

and, for *plane strain,*

$$E^1 = E(1-v)/[(1+v)\,(1-2v)]$$
$$\mu = v/(1-v)$$
$$\gamma = (1-2v)/[2\,(1-v)].$$

Hence, from (5.37) and (5.41),

$$[\tilde{k}] = \frac{E^1 x_2 y_3 t}{2} \begin{bmatrix} 0 & 0 & 0 & 0 & 0 & 0 \\ 0 & 1 & 0 & 0 & 0 & \mu \\ 0 & 0 & \gamma & 0 & \gamma & 0 \\ 0 & 0 & 0 & 0 & 0 & 0 \\ 0 & 0 & \gamma & 0 & \gamma & 0 \\ 0 & \mu & 0 & 0 & 0 & 1 \end{bmatrix}$$

and

$$
[k] = \frac{E^1 t}{2\,x_2 y_3}
\begin{bmatrix}
y_3{}^2 + \gamma x_{32}{}^2 & & \\
-(\gamma + \mu)y_3 x_{32} & \gamma y_3{}^3 + x_{32}{}^2 & \\
-y_3{}^2 - \gamma x_3 x_{32} & y_3(\gamma x_3 + \mu x_{32}) & \\
y_3(\gamma x_{32} + \mu x_3) & -\gamma y_3{}^2 - x_3 x_{32} & \\
\gamma x_2 x_{32} & -\gamma x_2 y_3 & \\
-\mu x_2 y_3 & x_2 x_{32} & \\
\end{bmatrix}
$$

$$
\begin{bmatrix}
y_3{}^2 + \gamma x_3{}^2 & \text{sym} & & \\
-x_3 y_3\,(\mu + \gamma) & \gamma y_3{}^2 + x_3{}^2 & & \\
-\gamma x_2 x_3 & \gamma x_2 y_3 & \gamma x_2{}^2 & \\
\mu x_2 y_3 & -x_2 x_3 & 0 & x_2{}^2 \\
\end{bmatrix} \quad (5.42)
$$

where

[k] = the elemental stiffness matrix for an in-plane three-node triangular element in *local coordinates*,

t = plate thickness

To obtain $[k^o]$, the elemental stiffness matrix for this element in *global coordinates* (Fig. 5.7) it will be necessary to use the expression

$$[k^o] = [\Xi]^\mathsf{T} [k] [\Xi]$$

where

$$
[\Xi] =
\begin{bmatrix}
\zeta & 0 & 0 \\
0 & \zeta & 0 \\
0 & 0 & \zeta \\
\end{bmatrix}
$$

$$
[\zeta] =
\begin{bmatrix}
c & s \\
-s & c \\
\end{bmatrix}
\text{(See Chapter 3)}
$$

$$
\begin{aligned}
c &= \cos \alpha = x^o{}_{21}/l_2 \\
s &= \sin \alpha = y^o{}_{21}/l_2 \\
x_{21}{}^o &= x_2{}^o - x_1{}^o \\
y_{21}{}^o &= y_2{}^o - y_1{}^o \\
l_2 &= \sqrt{[(x_{21}{}^o)^2 + (y_{21}{}^o)^2]}.
\end{aligned}
$$

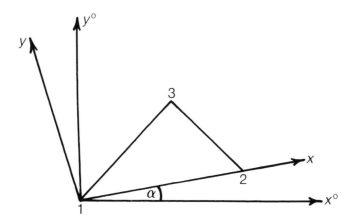

Fig. 5.7 – Triangle in global coordinates.

5.2.2   Example 5.1

Calculate the nodal deflections for the plate shown in Fig. 5.8, which is under plane stress.

$$E = 6.93 \times 10^4 \text{ MN/m}^2, v = 0.3, t = 2 \times 10^{-2} \text{ m.}$$

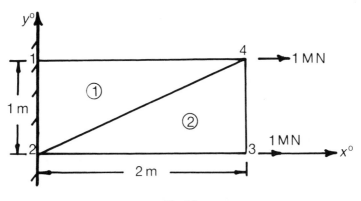

Fig. 5.8.

The nodes describing each element must be placed in an anticlockwise order. It is not necessary to start from the low node first. For example, for element 1, the nodes describing this element can be put in the following three combinations:

(a) 1–2–4, (b) 2–4–1, (c) 4–1–2.

These combinations will result in the systems of local coordinates shown in Figs. 5.9 to 5.11.

(a) 1–2–4

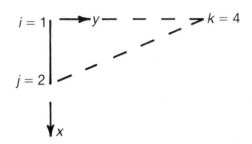

Fig. 5.9.

(b) 2–4–1

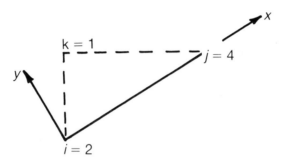

Fig. 5.10.

(c) 4–1–2

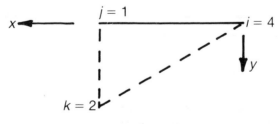

Fig. 5.11.

Thus it can be seen that if a different combination of i, j, and k nodes is used to describe the element, the elemental stiffness matrix will be different, although the system stiffness matrix will be the same. The reason for this is that although the $[k^o]$ matrix is different for each different combination of i–j–k, the stiffness influence coefficients corresponding to the appropriate nodal displacements will be the same. That is, there will simply be a redistribution of stiffness influence coefficients in $[k^o]$, as shown below.

(a) 1–2–4

$$[k^o_{1\text{-}2\text{-}4}] = \begin{array}{c} \begin{array}{cccccc} u_1{}^o & v_1{}^o & u_2{}^o & v_2{}^o & u_4{}^o & v_4{}^o \end{array} \\ \begin{bmatrix} a & b & c & d & e & f \\ b & g & h & i & j & k \\ c & h & l & m & n & o \\ d & i & m & p & q & r \\ e & j & n & q & s & t \\ f & k & o & r & t & u \end{bmatrix} \begin{array}{l} u_1{}^o \\ v_1{}^o \\ u_2{}^o \\ v_2{}^o \\ u_4{}^o \\ v_4{}^o . \end{array} \end{array}$$

(b) 2–4–1

$$[k^o_{2\text{-}4\text{-}1}] = \begin{array}{c} \begin{array}{cccccc} u_2{}^o & v_2{}^o & u_4{}^o & v_4{}^o & u_1{}^o & v_1{}^o \end{array} \\ \begin{bmatrix} l & m & n & o & c & h \\ & p & q & r & d & i \\ & & s & t & e & j \\ & & & u & f & k \\ \text{Symmetrical} & & & & a & b \\ & & & & & g \end{bmatrix} \begin{array}{l} u_2{}^o \\ v_2{}^o \\ u_4{}^o \\ v_4{}^o \\ u_1{}^o \\ v_1{}^o . \end{array} \end{array}$$

(c) 4–1–2

$$[k^o_{4\text{-}1\text{-}2}] = \begin{array}{c} \begin{array}{cccccc} u_4{}^o & v_4{}^o & u_1{}^o & v_1{}^o & u_2{}^o & v_2{}^o \end{array} \\ \begin{bmatrix} s & t & e & j & n & q \\ & u & f & k & o & r \\ & & a & b & c & d \\ & & & g & h & i \\ \text{Symmetrical} & & & & l & m \\ & & & & & p \end{bmatrix} \begin{array}{l} u_4{}^o \\ v_4{}^o \\ u_1{}^o \\ v_1{}^o \\ u_2{}^o \\ v_2{}^o . \end{array} \end{array}$$

For this particular example, element 1 will be described by nodes 1–2–4, and element 2 by nodes 2–3–4, so that the elemental stiffness matrices in global coordinates will be

$$[k_{(1)}°] =$$

	$u_1°$	$v_1°$	$u_2°$
	913.846		
	−495.000	1656.346	
	−533.077	266.539	533.077
	228.462	−1523.077	0.000
	−380.769	228.462	0.000
	266.538	−133.269	−266.538

	$v_2°$	$u_4°$	$v_4°$	
		sym		$u_1°$
				$v_1°$
				$u_2°$
	1523.077			$v_2°$
	− 228.462	380.769		$u_4°$
	0.000	0.000	133.269	$v_4°$

$$[k_{(2)}°] =$$

	$u_2°$	$v_2°$	$u_3°$
	380.769		
	0.000	133.269	
	−380.769	266.538	913.846
	228.462	−133.269	−495.000
	0.000	−266.539	−533.077
	−228.462	0.000	228.462

	$v_3°$	$u_4°$	$v_4°$	
		sym		$u_2°$
				$v_2°$
				$u_3°$
	1656.346			$v_3°$
	266.538	533.077		$u_4°$
	−1523.077	0.000	1523.077	$v_4°$.

Summing $[k_{(1)}°]$ and $[k_{(2)}°]$ and eliminating elements corresponding to nodes 1 and 2,

$$[K_{11}°] =$$

	$u_3°$	$v_3°$	$u_4°$	$v_4°$	
	913.846		sym		$u_3°$
	−495.000	1656.346			$v_3°$
	−533.077	266.538	913.846		$u_4°$
	228.462	−1523.077	0.000	1656.346	$v_4°$

and

$$[K_{11}^{\circ}]^{-1} = \begin{bmatrix} 0.002095 & & & \\ 0.001523 & 0.006722 & \text{sym} & \\ 0.000778 & -0.001072 & 0.001861 & \\ 0.001111 & 0.005972 & -0.001093 & 0.005942 \end{bmatrix}$$

Now,

$$\{q^{\circ}_F\} = \begin{Bmatrix} 1 \\ 0 \\ 1 \\ 0 \end{Bmatrix} \text{MN}$$

therefore

$$\{u_F\} = \begin{Bmatrix} u_3^{\circ} \\ v_3^{\circ} \\ u_4^{\circ} \\ v_4^{c} \end{Bmatrix} = \begin{Bmatrix} 0.2873 \\ 0.0451 \\ 0.2639 \\ 0.0018 \end{Bmatrix} \text{cm}$$

If 2 MN were assumed to be evenly distributed over the face of 3–4, then according to elementary elastic theory $u_3^{\circ} = u_4^{\circ} = 0.289$ cm, which compares with the finite element solution. It is interesting to note that in the finite element theory u_3° was greater than u_4°. This was because there were two elements at 4 and only one at 3.

5.2.3

The method of *minimum potential* is now used to obtain $[k^{\circ}]$ for an in-plane three-node triangular element. Consider the in-plane triangular element of Fig. 5.12 in global coordinates, and assume the following displacement functions:

$$u^{\circ} = \alpha_1 + \alpha_2 x^{\circ} + \alpha_3 y^{\circ}$$
$$v^{\circ} = \alpha_4 + \alpha_5 x^{\circ} + \alpha_6 y^{\circ} \tag{5.43}$$

@ $x^{\circ} = x_i^{\circ}$ and $y = y_i^{\circ}$
$$u_i^{\circ} = \alpha_1 + \alpha_2 x_i^{\circ} + \alpha_3 y_i^{\circ} \tag{5.44}$$

@ $x = x_j^{\circ}$ and $y = y_j^{\circ}$
$$u_j^{\circ} = \alpha_1 + \alpha_2 x_j^{\circ} + \alpha_3 y_j^{\circ} \tag{5.45}$$

Similarly, at $x = x_k^{\circ}$ and $y = y_k^{\circ}$,
$$u_k^{\circ} = \alpha_1 + \alpha_2 x_k^{\circ} + \alpha_3 y_k^{\circ} \tag{5.46}$$

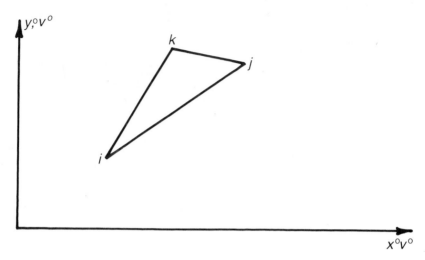

Fig. 5.12.

Solving (5.44) to (5.46) and substituting back into (5.43(a)), the following is
obtained:

$$u^{\circ} = \frac{1}{2\Delta} \left\lfloor (a_i + b_i x^{\circ} + c_i y^{\circ}) \, (a_j + b_j x^{\circ} + c_j y^{\circ}) \, (a_k + b_k x^{\circ} + c_k y^{\circ}) \right\rfloor \begin{Bmatrix} u_i^{\circ} \\ u_j^{\circ} \\ u_k^{\circ} \end{Bmatrix}$$

where

$$a_i = x_j^{\circ} y_k^{\circ} - x_k^{\circ} y_j^{\circ}$$
$$b_i = y_j^{\circ} - y_k^{\circ}$$
$$c_i = x_k^{\circ} - x_j^{\circ}$$
$$a_j = x_k^{\circ} y_i^{\circ} - x_i^{\circ} y_k^{\circ}$$
$$b_j = y_k^{\circ} - y_i^{\circ}$$
$$c_j = x_i^{\circ} - x_k^{\circ}$$
$$a_k = x_i^{\circ} y_j^{\circ} - x_j^{\circ} y_i^{\circ}$$
$$b_k = y_i^{\circ} - y_j^{\circ}$$
$$c_k = x_j^{\circ} - x_i^{\circ}.$$

$$\Delta = \text{area of triangle} = \frac{1}{2} \begin{vmatrix} 1 & x_i^{\circ} & y_i^{\circ} \\ 1 & x_j^{\circ} & y_j^{\circ} \\ 1 & x_k^{\circ} & y_k^{\circ} \end{vmatrix}.$$

By a similar process, an expression for 'v°' can be obtained, so that

$$\{u^o_{(x,y)}\} = \begin{Bmatrix} u^o \\ v^o \end{Bmatrix} = \begin{bmatrix} N_i & 0 & N_j & 0 & N_k & 0 \\ 0 & N_i & 0 & N_j & 0 & N_k \end{bmatrix} \begin{Bmatrix} u_i^o \\ v_i^o \\ u_j^o \\ v_j^o \\ u_k^o \\ v_k^o \end{Bmatrix}$$

$$= [N]\{u_i^o\}$$

$$N_i = (a_i + b_i x^o + c_i y^o)/2\Delta$$
$$N_j = (a_j + b_j x^o + c_j y^o)/2\Delta$$
$$N_k = (a_k + b_k x^o + c_k y^o)/2\Delta.$$

Now the coordinate strains are given by

$$\{\varepsilon^o\} = \begin{Bmatrix} \varepsilon_x^o \\ \varepsilon_y^o \\ \gamma_{x^o y^o} \end{Bmatrix} = \begin{Bmatrix} \dfrac{\partial u^o}{\partial x^o} \\[2mm] \dfrac{\partial v^o}{\partial y^o} \\[2mm] \dfrac{\partial u^o}{\partial y^o} + \dfrac{\partial v^o}{\partial x^o} \end{Bmatrix}$$

$$= \begin{bmatrix} \dfrac{\partial N_i}{\partial x^o} & 0 & \dfrac{\partial N_j}{\partial x^o} & 0 & \dfrac{\partial N_k}{\partial x^o} & 0 \\[2mm] 0 & \dfrac{\partial N_i}{\partial y^o} & 0 & \dfrac{\partial N_j}{\partial y^o} & 0 & \dfrac{\partial N_k}{\partial y^o} \\[2mm] \dfrac{\partial N_i}{\partial y^o} & \dfrac{\partial N_i}{\partial x^o} & \dfrac{\partial N_j}{\partial y^o} & \dfrac{\partial N_j}{\partial x^o} & \dfrac{\partial N_k}{\partial y^o} & \dfrac{\partial N_k}{\partial x^o} \end{bmatrix} \begin{Bmatrix} u_i^o \\ v_i^o \\ u_j^o \\ v_j^o \\ u_k^o \\ v_k^o \end{Bmatrix} \quad (5.47)$$

or

$$\{\varepsilon^o\} = [B]\{u_i^o\},$$

from which

$$[B] = \frac{1}{2\Delta} \begin{bmatrix} b_i & 0 & b_j & 0 & b_k & 0 \\ 0 & c_i & 0 & c_j & 0 & c_k \\ c_i & b_i & c_j & b_j & c_k & b_k \end{bmatrix}. \quad (5.48)$$

Hence, from (5.48) and (5.41), $[k^o]$ can be put in the following form:

$$[k^o] = t \begin{bmatrix} P_{ij} & Q_{ij} \\ \hline Q_{ji} & R_{ij} \end{bmatrix} \quad (5.49)$$

$$P_{ij} = 0.25\, E^1\, (b_i b_j + \mu\, c_i c_j)/\Delta$$
$$Q_{ij} = 0.25\, E^1\, (\gamma b_i c_j + \mu\, c_i b_j)/\Delta$$
$$Q_{ji} = 0.25\, E^1\, (\gamma b_j c_i + \mu\, c_j b_i)/\Delta$$
$$R_{ij} = 0.25\, E^1\, (c_i c_j + \mu\, b_i b_j)/\Delta.$$

This method can be seen to give a more satisfactory solution than that of the virtual work method for this case.

The main snag, however, with the in-plane element described in this section is that as it assumes a constant stress distribution across the element, errors occur when the element is used in areas of steep stress gradient. More sophisticated elements, which remedy this effect to some extent, are presented in Chapter 6.

5.2.4 In-plane annular element

Example 5.2
Determine the matrix of shape functions for the in-plane axisymmetric annular element of Fig. 5.13. Hence, or otherwise, obtain the elemental stiffness matrix for this element under plane stress and plane strain.

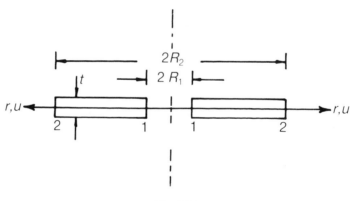

Fig. 5.13.

To obtain $[N]$
As this problem is axisymmetric, the radial deflection 'u' will be constant for any particular value of 'r'. Hence, for this case, it will be convenient to use two nodal circles to describe the element. Nodal circle 1 is the inner circumference, and nodal circle 2 is the outer circumference of this annular plate.

As there are two degrees of freedom, namely u_1 and u_2, it will be convenient to assume two unknowns for the radial deflection 'u', as follows:

$$\{u_{(x,y)}\} = u = \alpha_1 + \alpha_2 r. \tag{5.50}$$

The boundary conditions are

$$@\, r = R_1,\, u = u_1$$

and

$$@\, r = R_2,\, u = u_2.$$

Substituting these boundary conditions into (5.50),

$$u_1 = \alpha_1 + \alpha_2 R_1$$
$$u_2 = \alpha_1 + \alpha_2 R_2. \tag{5.51}$$

Solving (5.51),

$$\alpha_2 = \frac{(u_2 - u_1)}{(R_2 - R_1)}$$

$$\alpha_1 = \frac{(R_2 u_1 - R_1 u_2)}{(R_2 - R_1)}. \tag{5.52}$$

Substituting (5.52) into (5.50),

$$u = \frac{1}{(R_2 - R_1)} \left[(R_2 - r)\,(r - R_1) \right] \begin{Bmatrix} u_1 \\ u_2 \end{Bmatrix}$$

$$[N] = \frac{1}{(R_2 - R_1)} \left[(R_2 - r)\,(r - R_1) \right]$$

To obtain $[k]$

This is a two-dimensional problem, and from [30], it can be seen that the two coordinate strains in polar coordinates are given by:

$$\varepsilon_r = \text{radial strain} = \frac{du}{dr} = \lfloor -1/(R_2 - R_1)\ \ 1/(R_2 - R_1) \rfloor \{u_i\}$$

$$\varepsilon_\phi = \text{hoop or circumferential strain}$$

$$= \frac{u}{r} = \lfloor (R_2 - r)/[r\,(R_2 - R_1)]\ \ (r - R_1)/[r\,(R_2 - R_1)] \rfloor \{u_i\},$$

i.e.,

$$\{\varepsilon\} = \begin{Bmatrix} \varepsilon_r \\ \varepsilon_\phi \end{Bmatrix} = \frac{1}{(R_2 - R_1)} \begin{bmatrix} -1 & 1 \\ \dfrac{(R_2 - r)}{r} & \dfrac{(r - R_1)}{r} \end{bmatrix} \begin{Bmatrix} u_1 \\ u_2 \end{Bmatrix}$$

$$= [B] \{u_i\}, \tag{5.53}$$

To obtain $[D]$, the two-dimensional stress–strain relationships must be considered, which are

$$\{\sigma\} = \left\{ \begin{matrix} \sigma_r \\ \sigma_\phi \end{matrix} \right\} = E^1 \begin{bmatrix} 1 & \mu \\ \mu & 1 \end{bmatrix} \left\{ \begin{matrix} \varepsilon_r \\ \varepsilon_\phi \end{matrix} \right\} = [D]\{\varepsilon\}. \tag{5.54}$$

There is no shear stress in this plane, as both σ_r and σ_ϕ are principal stresses. Now,

$$[k] = \int [B]^T [D][B]\, d(\text{vol}),$$

and as this problem is axisymmetric,

$$d(\text{vol}) = 2\pi r\, dr \cdot t$$

where

$r = $ any radius, $t = $ plate thickness.

Therefore

$$[k] = \int_{R_1}^{R_2} [B]^T [D][B]\, 2\pi r\, drt. \tag{5.55}$$

Substituting (5.53) and (5.54) into (5.55), the following is obtained for the elements of the elemental stiffness matrix for an in-plane annular plate under plane stress [33].

$$k_{11} = CN\{R_2\,[R_2\, ln\,(R_2/R_1) - 2\,(1+\mu)\,(R_2 - R_1)]$$
$$+ (1+\mu)\,(R_2{}^2 - R_1{}^2)\}$$
$$k_{12} = k_{21} = CN\,\{- R_1 R_2\, ln\,(R_2/R_1)\}$$
$$k_{22} = CN\,\{R_1\,[R_1\, ln\,(R_2/R_1) - 2\,(1+\mu)\,(R_2 - R_1)]$$
$$+ (1+\mu)\,(R_2{}^2 - R_1{}^2)\}$$
$$CN = 2\pi\,E^1\,t/(R_2 - R_1)^2.$$

5.3 BEAM ELEMENT

Example 5.3
Determine the matrix of shape functions for the beam element of Fig. 5.14. Hence, or otherwise, determine the stiffness matrix for this element.

To determine $[N]$
It will be convenient to assume a displacement function 'v' with four unknowns, as there are four boundary conditions, namely v_1, Θ_1, v_2 and Θ_2.

$$\{u_{(x,\,y)}\} = v = \alpha_1 + \alpha_2 x + \alpha_3 x^2 + \alpha_4 x^3. \tag{5.56}$$

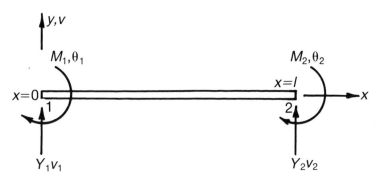

Fig. 5.14 – Beam element.

The displacement 'θ' must not be separately assumed, as 'θ' is a function of 'v', as follows:

$$\theta = -\frac{dv}{dx} = -\alpha_2 - 2\alpha_3 x - 3\alpha_4 x^2. \tag{5.57}$$

The four boundary conditions which are required to solve the four unknowns are

$$x = 0, v = v_1 \text{ and } \theta = \theta_1$$
$$x = l, v = v_2 \text{ and } \theta = \theta_2.$$

From the first two boundary conditions,

$$\alpha_1 = v_1 \tag{5.58}$$

and

$$\alpha_2 = -\theta_1. \tag{5.59}$$

From the third and fourth boundary conditions,

$$\alpha_3 = \frac{3}{l^2}(v_2 - v_1) + \frac{1}{l}(2\theta_1 + \theta_2) \tag{5.60}$$

and

$$\alpha_4 = \frac{2}{l^3}(v_1 - v_2) - \frac{1}{l^2}(\theta_1 + \theta_2). \tag{5.61}$$

Substituting (5.58) to (5.61) into (5.56):

$$v = v_1 - \Theta_1 x + \left[\frac{3}{l^2}(v_2 - v_1) + \frac{1}{l}(2\theta_1 + \theta_2)\right]x^2$$

$$+ \left[\frac{2}{l^3}(v_1 - v_2) - \frac{1}{l^2}(\theta_1 + \theta_2)\right]x^3. \tag{5.62}$$

Putting $\xi = x/l$, and rearranging (5.62),

$$v = [(1 - 3\ \xi^2 + 2\xi^3)\ l\ (-\xi + 2\xi^2 - \xi^3)\ (3\xi^2 - 2\xi^3)\ l\ (\xi^2 - \xi^3)] \begin{Bmatrix} v_1 \\ \theta_1 \\ v_2 \\ \theta_2 \end{Bmatrix}$$

(5.63)

$$= [N]\ \{u_i\}$$

where

$$[N] = [(1 - 3\ \xi^2 + 2\xi^3)l(-\xi + 2\xi^2 - \xi^3)\ (3\xi^2 - 2\xi^3)\ l\ (\xi^2 - \xi^3)]. \quad (5.64)$$

Equations (5.62) and (5.63) can be recognised as a Hermite polynomial [34], which has deflection and slope continuity at the nodes (i.e., the polynomial collocates and osculates to the first degree).

To determine $[k]$
From the elementary theory of beams [35], the strain energy in bending is given by the expression

$$U_e = \int \frac{M^2}{2EI}\ dx.$$

(5.65)

Now,

$$EI\ \frac{d^2v}{dx^2} = M,$$

(5.66)

Therefore

$$U_e = \frac{1}{2} \int EI \left(\frac{d^2v}{dx^2} \right)^2 dx.$$

(5.67)

In matrix form, equation (5.67) appears as

$$U_e = \frac{1}{2} \int \left\{ \frac{d^2v}{dx^2} \right\}^T [EI] \left\{ \frac{d^2v}{dx^2} \right\} dx.$$

(5.68)

Equation (5.68) must be written in the form shown, because U_e is a scalar. On comparing (5.66) with (5.12) and (5.68) with (5.19), it can be seen that for beams in bending,

$$\{\sigma\} = M$$

$$\{\varepsilon\} = \frac{d^2v}{dx^2}$$

$$[D] = EI$$

d(vol) ≡ dx (this is because the cross-sectional area is already incorporated in 'I').

E = elastic modulus

I = 2nd moment of area.

Now,

$$\{\varepsilon\} = \frac{d^2v}{dx^2} = \frac{d^2v}{l^2 d\xi^2}$$

$$= \frac{1}{l^2}\left[(-6+12\xi)\, l\,(4-6\xi)\,(6-12\xi)\, l\,(4-6\xi)\right]\begin{Bmatrix} v_1 \\ \theta_1 \\ v_2 \\ \theta_2 \end{Bmatrix}. \qquad (5.69)$$

Comparing (5.69) with (5.25), it can be seen that

$$[B] = \frac{1}{l^2}\left[(-6+12\xi)\, l\,(4-6\xi)\,(6-12\xi)\, l\,(4-6\xi)\right]$$

Now,

$$[k] = \int [B]^T [D] [B]\, dx$$

$$= \int_0^1 [B]^T [D] [B]\, l d\xi$$

$$= \frac{EI}{l^3}\int_0^1 \begin{bmatrix} (-6+12\xi) \\ l\,(4-6\xi) \\ (6-12\xi) \\ l\,(4-6\xi) \end{bmatrix}\left[(-6+12\xi)\, l\,(4-6\xi)\,(6-12\xi)\, l\,(4-6\xi)\right] l d\xi.$$

$$\qquad (5.70)$$

Multiplying out (5.70) and integrating, it can be shown that the matrix of equation (4.12) can be obtained.

5.4 PLATE BENDING PROBLEMS

5.4.1 Non-conforming triangular element

A stiffness matrix will be derived for the triangular element shown in Fig. 5.15, by the principle of virtual work. The stiffness matrix is due to Clough & Tocher [36], and although a useful one, not all the continuity requirements are satisfied for the assumed displacement function. For example, although deflections and slopes are continuous at the nodal points, they are not necessarily so along the boundaries of the element. This can be readily observed if the slope θ_x is considered along the boundary 1–2. Now,

$$\theta_x = \partial w / \partial y$$

where w = lateral deflection.

Hence, from (5.71),

$$\theta_{x1-2} = \alpha_3 + \alpha_5 x + \alpha_8 x^2.$$

As, however, so far as conditions for slopes along the x axis are concerned, because only θ_{x1} and θ_{x2} are expressed at nodes 1 and 2, it will not be possible to ensure slope continuity along the boundary 1–2 in the above expression. This is because there is a parabolic variation in slope with only two boundary conditions for θ_x. For this reason, this element is called a *non-conforming* one. It is evident from this that, as there is no slope continuity there will be no curvature continuity, and also that, for an element to be fully compatible it will be necessary to satisfy both these conditions in addition to deflection continuity. Nevertheless, the resulting expressions can be extremely useful, especially for shell analysis.

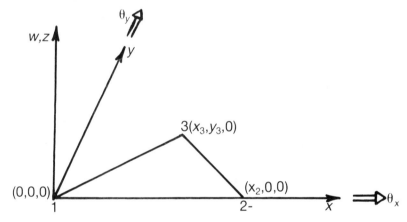

Fig. 5.15 – Flat triangular element in bending.

As there are nine nodal displacements, it will be convenient to assume a displacement function with nine constants, as follows:

$$w = \alpha_1 + \alpha_2 x + \alpha_3 y + \alpha_4 x^2 + \alpha_5 xy + \alpha_6 y^2 + \alpha_7 x^3$$
$$+ \alpha_8 (x^2 y + xy^2) + \alpha_9 y^3. \qquad (5.71)$$

It can be seen from equation (5.71) that to express the complete cubic polynomial, it is necessary to use ten coefficients; however, as there are only nine degrees of freedom per element, only nine coefficients can be used. The reason for this is that, as there are only nine degrees of freedom per element, only nine simultaneous equations can be obtained, hence, only nine unknown 'α's' can be determined in terms of $\{u_i\}$, that is, $[C]$ must be square.

For this particular case, it will be found convenient to join the $x^2 y$ and xy^2 terms together. It is, of course, possible to treat other terms in a similar manner to these, but Clough & Tocher {36}, considered this combination to be the most satisfactory.

The vector of nodal displacements is given by

$$\{u_i\} = \{w_1 \ \theta_{x1} \ \theta_{y1} \ w_2 \ \theta_{x2} \ \theta_{y2} \ w_3 \ \theta_{x3} \ \theta_{y3}\}$$
$$= [C]\{\alpha\}.$$

Hence, by substituting the following nine boundary conditions into $\{u_i\}$,

@ $x = y = 0$, $w = w_1$, $\theta_x = \theta_{x1}$ and $\theta_y = \theta_{y1}$

@ $x = x_2$ and $y = 0$, $w = w_2$, $\theta_x = \theta_{x2}$ and $\theta_y = \theta_{y2}$

@ $x = x_3$ and $y = y_3$, $w = w_3$, $\theta_x = \theta_{x3}$ and $\theta_y = \theta_{y3}$

the matrix $[C]$ can be determined, as shown in equation (5.72).

$$[C] = \begin{bmatrix} 1 & 0 & 0 & 0 & 0 & 0 & 0 & 0 & 0 \\ 0 & 0 & 1 & 0 & 0 & 0 & 0 & 0 & 0 \\ 0 & -1 & 0 & 0 & 0 & 0 & 0 & 0 & 0 \\ 1 & x_2 & 0 & x_2^2 & 0 & 0 & x_2^3 & 0 & 0 \\ 0 & 0 & 1 & 0 & x_2 & 0 & 0 & x_2^2 & 0 \\ 0 & -1 & 0 & -2x_2 & 0 & 0 & -3x_2^2 & 0 & 0 \\ 1 & x_3 & y_3 & x_3^2 & x_3 y_3 & y_3^2 & x_3^3 & (x_3^2 y_3 + x_3 y_3^2) & y_3^3 \\ 0 & 0 & 1 & 0 & x_3 & 2y_3 & 0 & (x_3^2 + 2x_3 y_3) & 3y_3^2 \\ 0 & -1 & 0 & -2x_3 & -y_3 & 0 & -3x_3^2 & -(2x_3 y_3 + y_3^2) & 0 \end{bmatrix}$$

$$(5.72)$$

where,

$$\theta_x = \frac{\partial w}{\partial y}$$

and

$$\theta_y = -\frac{\partial w}{\partial x}.$$

Now, for plate bending problems it can be shown that internal virtual work done by a bending moment is equal to the bending moment × virtual curvature. That is, if

$$\{\sigma\} = \begin{Bmatrix} M_x \\ M_y \\ M_{xy} \end{Bmatrix} = \begin{Bmatrix} \text{Bending moment corresponding to } \theta_x \\ \text{Bending moment corresponding to } \theta_y \\ \text{Twisting moment in } x\text{–}y \text{ plane} \end{Bmatrix}$$

and

$$\{\bar{\varepsilon}\} = \left\{ \begin{array}{c} \dfrac{\partial^2 \bar{w}}{\partial y^2} \\[2ex] -\dfrac{\partial^2 \bar{w}}{\partial x^2} \\[2ex] \dfrac{2\partial^2 \bar{w}}{\partial x\, \partial y} \end{array} \right\} = \left\{ \begin{array}{l} \text{Curvature corresponding to } \theta_x \\[2ex] \text{Curvature corresponding to } \theta_y \\[2ex] \text{Twisting curvature in } x\text{–}y \text{ plane} \end{array} \right\},$$

then, virtual work done $= \{\bar{\varepsilon}\}^{\mathrm{T}} \{\sigma\}$.

The bending curvature relationships [30] are,

$$M_x = D\left(\frac{\partial^2 w}{\partial y^2} + v\, \frac{\partial^2 w}{\partial x^2} \right)$$

$$M_y = -D\left(\frac{\partial^2 w}{\partial x^2} + v\, \frac{\partial^2 w}{\partial y^2} \right)$$

$$M_{xy} = D(1-v)\, \frac{\partial^2 w}{\partial x\, \partial y}$$

where

$$D = \frac{Et^3}{12(1-v^2)} = \text{Flexural rigidity,}$$

or in matrix form,

$$\left\{ \begin{array}{c} M_x \\ M_y \\ M_{xy} \end{array} \right\} = D \begin{bmatrix} 1 & -v & 0 \\ -v & 1 & 0 \\ 0 & 0 & \dfrac{(1-v)}{2} \end{bmatrix} \left\{ \begin{array}{c} \dfrac{\partial^2 w}{\partial y^2} \\[2ex] -\dfrac{\partial^2 w}{\partial x^2} \\[2ex] \dfrac{2\partial^2 w}{\partial x\, \partial y} \end{array} \right\}, \tag{5.73}$$

that is,

$$\{\sigma\} = [D]\,\{\varepsilon\}.$$

Now, as

$$\{\varepsilon\} = \left\{ \begin{array}{c} \dfrac{\partial^2 w}{\partial y^2} \\[2ex] -\dfrac{\partial^2 w}{\partial x^2} \\[2ex] 2\dfrac{\partial^2 w}{\partial x\, \partial y} \end{array} \right\} = [B_{(x,\, y)}]\,\{\alpha\}$$

$$[B_{(x,\,y)}] = \begin{bmatrix} 0 & 0 & 0 & 0 & 0 & 2 & 0 & 2x & 6y \\ 0 & 0 & 0 & -2 & 0 & 0 & -6x & -2y & 0 \\ 0 & 0 & 0 & 0 & 2 & 0 & 0 & 4(x+y) & 0 \end{bmatrix}. \tag{5.74}$$

Substituting (5.73) and (5.74) into (5.17),

$$[\tilde{k}] = \int_A D \begin{bmatrix} 0 & 0 & 0 & 0 & 0 & 0 \\ 0 & 0 & 0 & 0 & 0 & 0 \\ 0 & 0 & 0 & 0 & 0 & 0 \\ 0 & 0 & 0 & 4 & 0 & 4v \\ 0 & 0 & 0 & 0 & 2(1-v) & 0 \\ 0 & 0 & 0 & 4v & 0 & 4 \\ 0 & 0 & 0 & 12x & 0 & 12vx \\ 0 & 0 & 0 & 4(vx+y) & 4(1-v)(x+y) & 4(x+vy) \\ 0 & 0 & 0 & 12vy & 0 & 12y \end{bmatrix}$$

$$\begin{matrix} 0 & 0 & 0 \\ 0 & 0 & 0 \\ 0 & 0 & 0 \\ 12x & 4(vx+y) & 12vy \\ 0 & 4(1-v)(x+y) & 0 \\ 12vx & 4(x+vy) & 12y \\ 36x^2 & 12x(vx+y) & 36vxy \\ 12x(vx+y) & 4\{(3-2v)(x^2+y^2) & 12y(x+vy) \\ & +2(2-v)xy\} & \\ 36vxy & 12y(x+vy) & 36y^2 \end{matrix} \; dA \tag{5.75}$$

where,

$$\iint dx\, dy = \tfrac{1}{2} x_2 y_3 = \Delta$$
$$\iint y\, dx\, dy = \tfrac{1}{6} y_3^2 x_2 = \Delta y_3/3$$
$$\iint y^2\, dx\, dy = \tfrac{1}{12} y_3^3 x_2 = \Delta y_3^2/6$$
$$\iint x\, dy\, dx = \tfrac{1}{6} y_3 x_2 (x_2 + x_3) = \Delta(x_2 + x_3)/3$$
$$\iint x^2\, dx\, dy = \tfrac{1}{12} y_3 x_2 (x_2^2 + x_2 x_3 + x_3^2)$$
$$\qquad\qquad = \Delta(x_2^2 + x_2 x_3 + x_3^2)/6$$
$$\iint xy\, dx\, dy = \tfrac{1}{24} y_3^2 x_2 (x_2 + 2x_3) = \Delta y_3 (x_2 + 2x_3)/12.$$

Deficiencies of [C]
Problems occur in inverting the matrix of equation (5.72), particularly if the '*i*' node is on a right-angle, as shown in Fig. 5.16. This is because for some such cases, [C] is ill-conditioned (or singular), and attempting to obtain its inverse, is either impossible or results in numerical instability.

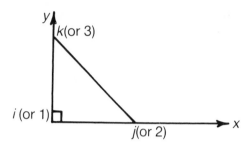

Fig. 5.16.

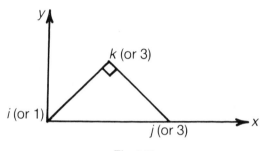

Fig. 5.17.

Whenever $[C]$ is ill-conditioned, the problem of inverting it can be overcome by choosing a more suitable system of local axes, as shown in Fig. 5.17.

5.4.2 Other plate bending elements

One of the problems with the element of Clough & Tocher is that it does not always converge to the correct result, with an increase in the refinement of the mesh.

Another more serious problem with it, is that $[C]$ is sometimes singular, and a method of overcoming this deficiency is to use area coordinates, as described in reference [47].

Other popular plate bending elements take the form of rectangles and quadrilaterals, and one of these is described in Chapter 7.

5.4.3 Large deflections of plates

It must be emphasised here that the analysis described in this section is only suitable for small lateral deflections where, in general, the maximum lateral deflection is not greater than half the thickness of the plate.

For large deflections of plates, the geometry of the plate changes, so that the plate becomes a shallow shell, resisting the bulk of its load in a membrane manner.

To analyse the large deflections of plates by finite elements is difficult and inefficient, and for such cases, it is better to resort to other methods, wherever possible.

5.5 CURVED SHELLS

A very useful method of representing thin curved shells is by the use of flat plate elements, which when joined together, can make an approximate description of quite complicated shells, as shown in Figs 5.18 and 5.19. Triangular and quadrilateral elements are particularly powerful, as they can represent doubly curved shells of varying thickness with cutouts and other discontinuities, but rectangular elements are only suitable for cylindrical shells, folded slabs, etc.

For shell analysis, both in-plane and out-of-plane effects are important, but these can be quite easily catered for by superimposing the in-plane stiffness matrix for a plate with that of its bending stiffness. Assuming the element has corner nodes only, the resulting 'shell' stiffness matrix will be of order 18×18 for the triangular element and 24×24 for the quadrilateral and rectangular elements. An alternative concept of these matrices is that they are 3×3 and 4×4 supermatrices respectively, the elements of which are the following 6×6 matrices, where the subscripts p and b denote in-plane and bending effects respectively.

$$[k_{ij}] = \begin{bmatrix} k_p^{xx} & k_p^{xy} & 0 & 0 & 0 & 0 \\ k_p^{yx} & k_p^{yy} & 0 & 0 & 0 & 0 \\ 0 & 0 & k_b^{zz} & k_b^{zx} & k_b^{zy} & 0 \\ 0 & 0 & k_b^{xz} & k_b^{xx} & k_b^{xy} & 0 \\ 0 & 0 & k_b^{yz} & k_b^{yx} & k_b^{yy} & 0 \\ 0 & 0 & 0 & 0 & 0 & 0 \end{bmatrix} \begin{matrix} u \\ v \\ w \\ \theta_x \\ \theta_y \\ \theta_z . \end{matrix}$$

$$\begin{matrix} u & v & w & \theta_x & \theta_y & \theta_z \end{matrix}$$

The six columns and rows of this matrix correspond to the six degrees of freedom at each node $- u, v, w, \theta_x, \theta_y$ and θ_z (see Fig. 5.20); and x, y, and z are the local coordinates. It should be noted here that the last column and row are set to zero, because as the element is in the x–y plane, θ_z has no value.

It is, of course, necessary to transform these stiffness matrices from local to global coordinates, and this can be carried out by the use of (3.31) and (3.14), the order of $[\Xi]$ being the same as that of $[k]$.

$N.B.$ It should be noted that if a coplanar section is met in the analysis of a doubly curved shell, the computer program can fail because of the rows and columns of zeros that will occur with respect to θ_z. Zienkiewicz [13], however, has shown how this deficiency can be overcome by making an approximation for these stiffness influence coefficients.

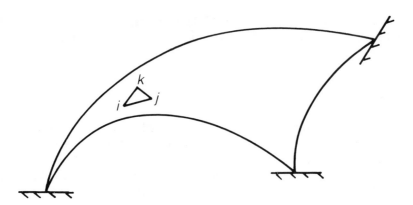

Fig. 5.18 – Doubly curved shell.

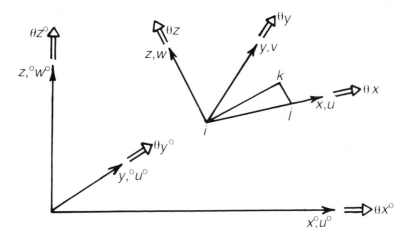

Fig. 5.19 – Cylindrical shells.

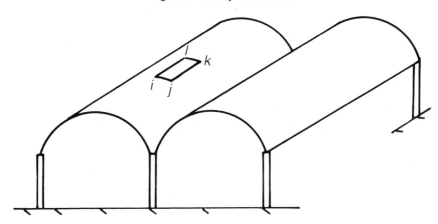

Fig. 5.20 – Flat shell element.

Apart from the difficulty of using this element to represent coplanar sections, the suitability of the element to model steep shells and areas of sudden discontinuity leaves much in question. This is because the element is not doubly curved, and researchers have found that is a serious deficiency in many cases.

As a result of this requirement, many curved elements have been produced in recent years [37, 38]. Some of these elements are in the form of cylindrical shells, and others are doubly curved triangles, as shown in Figs 5.21 and 5.22. To describe these elements in detail is beyond the scope of this book, and the reader is referred to more advanced books on this topic [13, 19, 39].

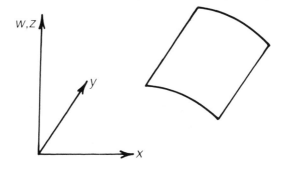

Fig. 5.21 – Cylindrical shell element.

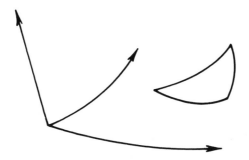

Fig. 5.22 – Doubly-curved triangular element.

5.6 INITIAL STRAINS AND THERMAL EFFECTS

Problems involving thermal stresses have presented much analytical challenge in the past, but today, with the aid of the finite element method, many of these difficulties can be overcome.

Let

$\{\varepsilon_0\}$ = a vector of initial strains, (possibly due to thermal effects), then (5.12) becomes

$$\{\sigma\} = [D](\{\varepsilon\} - \{\varepsilon_0\}). \tag{5.76}$$

Let

$\{\bar{\varepsilon}\}$ = a vector of virtual strains

$\{\sigma_0\} = [D]\{\varepsilon_0\}$

$[D]$ = a matrix of elastic constants.

Virtual work done due to initial strains,

$$= \int \{\bar{\varepsilon}\}^T \{\sigma_0\} \, d(\text{vol})$$
$$= \int \{\bar{\varepsilon}\}^T [D] \{\varepsilon_0\} \, d(\text{vol})$$
$$= \{\bar{u}_i\}^T [C^{-1}]^T \int [B_{(x, y)}]^T [D] \{\varepsilon_0\} \, d(\text{vol}). \tag{5.77}$$

Let

$\{p_T\}$ = a vector of nodal forces due to initial strains.

Then virtual work done by $\{P_T\}$

$$= \{\bar{u}_i\}^T \{P_T\}. \tag{5.78}$$

Equating (5.77) to (5.78),

$$\{P_T\} = [C^{-1}]^T \int_{\text{vol}} [B_{(x, y)}]^T [D] \{\varepsilon_0\} \, d(\text{vol}), \tag{5.79}$$

so that

$$\{q\} = \sum (\{P_i\} + \{P_T\}). \tag{5.80}$$

Now from (5.25),

$$[B] = [B_{(x, y)}][C^{-1}]$$

therefore

$\{P_T\}$ can be put in the alternative form

$$\{P_T\} = \int_{\text{vol}} [B]^T [D] \{\varepsilon_0\} \, d(\text{vol}). \tag{5.81}$$

5.7 TO CALCULATE ELEMENTAL NODAL FORCES DUE TO THERMAL EFFECTS

Equivalent load vectors $\{P_T\}$ due to thermal effects, will be calculated for various one- and two-dimensional elements.

5.7.1 Rod element

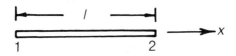

Fig. 5.23.

Let,

T = temperature rise

α = coefficient of linear expansion

E = elastic modulus.

From (5.79),

$$\{P_T\} = \int [B]^T [D] \{\varepsilon_0\} \, d(\text{vol}).$$

Substituting (5.13) and (5.27) into the above,

$$\{P_T\} = \frac{1}{l} \int \begin{bmatrix} -1 \\ 1 \end{bmatrix} E\alpha T \, A dx$$

$$= E\alpha T A \begin{Bmatrix} -1 \\ 1 \end{Bmatrix} \begin{matrix} u_1 \\ u_2 \end{matrix}.$$

5.7.2

For a *two-dimensional rod element* in global coordinates,

$$\{P_T\} = [\Xi]^T E\alpha T A \begin{Bmatrix} -1 \\ 0 \\ 1 \\ 0 \end{Bmatrix} \begin{matrix} u_1 \\ v_1 \\ u_2 \\ v_2 \end{matrix}. \tag{5.82}$$

Substituting $[\Xi]$ from (3.7) into (5.82),

$$\{P_T\} = E\alpha T A \left[\begin{array}{cc|cc} c & s & \\ -s & c & & 0_2 \\ \hline & 0_2 & c & s \\ & & -s & c \end{array} \right] \begin{Bmatrix} -1 \\ 0 \\ -1 \\ 0 \end{Bmatrix}$$

$$= E\alpha T A \begin{Bmatrix} -c \\ -s \\ c \\ s \end{Bmatrix}. \tag{5.83}$$

5.7.3

For a *three-dimensional rod* in global coordinates,

$$\{P_T\} = [\Xi]^T E\alpha T A \begin{Bmatrix} -1 \\ 0 \\ 0 \\ 1 \\ 0 \\ 0 \end{Bmatrix} \begin{matrix} u_1 \\ v_1 \\ w_1 \\ u_2 \\ v_2 \\ w_2 \end{matrix} \quad (5.84)$$

Substituting $[\Xi]$ from (3.30) into (5.84),

$$\{P_T\} = E\alpha T A \left[\begin{array}{ccc|ccc} C_{x,x^o} & C_{y,x^o} & C_{z,x^o} & & & \\ C_{x,y^o} & C_{y,y^o} & C_{z,y^o} & & 0_3 & \\ C_{x,z^o} & C_{y,z^o} & C_{z,z^o} & & & \\ \hline & & & C_{x,x^o} & C_{y,x^o} & C_{z,x^o} \\ & 0_3 & & C_{x,y^o} & C_{y,y^o} & C_{z,y^o} \\ & & & C_{x,z^o} & C_{y,z^o} & C_{z,z^o} \end{array} \right] \begin{Bmatrix} -1 \\ 0 \\ 0 \\ 1 \\ 0 \\ 0 \end{Bmatrix}$$

$$= E\alpha T A \begin{Bmatrix} -C_{x,x^o} \\ -C_{x,y^o} \\ -C_{x,z^o} \\ C_{x,x^o} \\ C_{x,y^o} \\ C_{x,z^o} \end{Bmatrix} \begin{matrix} u_1^o \\ v_1^o \\ W_1^o \\ u_2^o \\ v_2^o \\ w_2^o \end{matrix}$$

5.7.4

Similarly for a *two-dimensional rigid-jointed frame*,

$$\{P_T\} = E\alpha T A \begin{Bmatrix} -c \\ -s \\ 0 \\ c \\ s \\ 0 \end{Bmatrix}$$

and a *three-dimensional rigid-jointed frame*,

$$\{P_T\} = E\alpha T A \begin{Bmatrix} -C_{x,x^o} \\ -C_{x,y^o} \\ -C_{x,z^o} \\ 0 \\ 0 \\ 0 \\ C_{x,x^o} \\ C_{x,y^o} \\ C_{x,z^o} \\ 0 \\ 0 \\ 0 \end{Bmatrix}.$$

5.7.5

For an *in-plane plate*, in global coordinates,

$$\{P_T\} = \int [B]^T [D] \{\varepsilon_0\} \, d(\text{vol})$$

$$= \frac{t}{2\Delta} \int\!\!\int \begin{bmatrix} b_i & 0 & c_i \\ 0 & c_i & b_i \\ b_j & 0 & c_j \\ 0 & c_j & b_j \\ b_k & 0 & c_k \\ 0 & c_k & b_k \end{bmatrix} E^1 \begin{bmatrix} 1 & \mu & 0 \\ \mu & 1 & 0 \\ 0 & 0 & \gamma \end{bmatrix} \begin{Bmatrix} \alpha T \\ \alpha T \\ 0 \end{Bmatrix} dx\,dy$$

$$= \frac{E^1 t}{2\Delta} \int\!\!\int \begin{bmatrix} b_i & 0 & c_i \\ 0 & c_i & b_i \\ b_j & 0 & c_j \\ 0 & c_j & b_j \\ b_k & 0 & c_k \\ 0 & c_k & b_k \end{bmatrix} \begin{bmatrix} (1+\mu)\alpha T \\ (1+\mu)\alpha T \\ 0 \end{bmatrix} dx\,dy$$

$$= \frac{E^1 t(1+\mu)\alpha T}{2} \begin{Bmatrix} b_i \\ c_i \\ b_j \\ c_j \\ b_k \\ c_k \end{Bmatrix} \text{ for } \textit{plane stress}$$

and

$$\{P_\mathrm{T}\} = \frac{E^1(1+\mu)(1+v)\alpha T}{2} \begin{Bmatrix} b_i \\ c_i \\ b_j \\ c_j \\ b_k \\ c_k \end{Bmatrix} \text{ for } plane\ strain,$$

where b_i, c_i, etc., are defined in section 5.2.3.

as

$$\{\varepsilon\} = (1+v)\alpha T \begin{Bmatrix} 1 \\ 1 \\ 0 \end{Bmatrix}.$$

N.B. It must be remembered that (5.76) must be used for calculating 'stresses'.

5.8   DISTRIBUTED LOADS

A method is now given of calculating a vector of nodal loads $\{P_\mathrm{w}\}$ due to the effects of distributed loading.

Let $\{P_{(x,\ y)}\}$ = a vector of distributed loads which may be acting in the x, y, and z directions, depending on whether the load is one-, two- or three-dimensional. If $\{\bar{u}_{(x,\ y)}\}$ = a vector of virtual displacements, then the virtual work done by the distributed load

$$= \int_\mathrm{vol} \{\bar{u}_{(x,\ y)}\}^\mathrm{T} \{P_{(x,\ y)}\} \,\mathrm{d}(\mathrm{vol})$$
$$= \int \{\bar{\alpha}\}^\mathrm{T} [M_{(x,\ y)}]^\mathrm{T} \{P_{(x,\ y)}\} \,\mathrm{d}(\mathrm{vol})$$
$$= \{\bar{u}_i\}^\mathrm{T} [C^{-1}]^\mathrm{T} \int [M_{(x,\ y)}]^\mathrm{T} \{P_{(x,\ y)}\} \,\mathrm{d}(\mathrm{vol}) \qquad (5.85)$$

Let $\{P_\mathrm{w}\}$ = a vector of nodal loads equivalent to the distributed load. The virtual work done by $\{P_\mathrm{w}\}$

$$= \{\bar{u}_i\}^\mathrm{T} \{P_\mathrm{w}\} \qquad (5.86)$$

Equating (5.85) and (5.86),

$$\{P_\mathrm{w}\} = [C^{-1}]^\mathrm{T} \int_\mathrm{vol} [M_{(x,\ y)}]^\mathrm{T} \{P_{(x,\ y)}\} \,\mathrm{d}(\mathrm{vol}). \qquad (5.87)$$

An alternative expression to (5.87) can be obtained by using the relation

$$[N] = [M_{(x,\ y)}][C^{-1}]$$

therefore

$$\{P_w\} = \int [N]^T \{P_{(x,\,y)}\} \, d(vol).$$ (5.88)

To determine $\{P_w\}$ for a beam element under a uniformly distributed load 'w',

$$\{P_w\} = \int_0^1 \begin{Bmatrix} (1 - 3\xi^2 + 2\xi^3) \\ l(-\xi + 2\xi^2 - \xi^3) \\ (3\xi^2 - 2\xi^3) \\ l(2\xi^2 - \xi^3) \end{Bmatrix} wl\,d\xi$$

$$= \begin{Bmatrix} wl/2 \\ -wl^2/12 \\ wl/2 \\ wl^2/12 \end{Bmatrix}$$

which can be seen to be the negative resultants, (see Fig. 5.24) of the end fixing forces for element 1–2, as used and de   ed in Chapter 4.

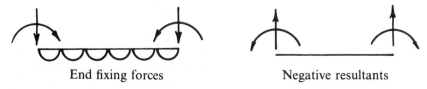

End fixing forces   Negative resultants

Fig. 5.24.

N.B. It should be noted that after $\{\sigma\}$ has been calculated, the end fixing 'forces' must be taken away, as in Chapter 4.

5.9   REDUCTION OF STIFFNESS MATRIX

This is a useful process for reducing the size of large problems, especially at nodes where no external loads exist. Turner *et al* [5] have shown that, when a structure has no external nodes at certain nodal degrees of freedom, the equations can be rearranged so that all displacements corresponding to zero loads can be kept together, as follows:

$$\begin{Bmatrix} q_a \\ 0 \end{Bmatrix} \begin{bmatrix} K_{aa} & K_{ab} \\ K_{ba} & K_{bb} \end{bmatrix} \begin{Bmatrix} u_a \\ u_b \end{Bmatrix}.$$

Expanding the above,

$$\{q_a\} = [K_{aa}]\{u_a\} + [K_{ab}]\{u_b\}$$ (5.89)

and

$$\{0\} = [K_{ba}]\{u_a\} + [K_{bb}]\{u_b\}$$

or

$$\{u_b\} = -[K_{bb}]^{-1}[K_{ba}]\{u_a\}. \tag{5.90}$$

Substituting (5.90) into (5.89),

$$\begin{aligned}
\{q_a\} &= [K_{aa}]\{u_a\} - [K_{ab}][K_{bb}]^{-1}[K_{ba}]\{u_a\} \\
&= ([K_{aa}] - [K_{ab}][K_{bb}]^{-1}[K_{ba}])\{u_a\} \tag{5.91} \\
&= [K_c]\{u_a\},
\end{aligned}$$

where

$[K_c]$ = condensed stiffness matrix

$$= [K_{aa}] - [K_{ab}][K_{bb}]^{-1}[K_{ba}]. \tag{5.92}$$

From (5.92), it can be seen that if there are only a few nodal loads on a large structure, the size of the stiffness matrix can be considerably reduced.

5.9.1

Irons [40] has shown how equation (5.92) can be put in the form of a continuous reduction technique, where displacements with loads and other required displacements are called 'masters', whilst displacements without loads, which are not required, are called 'slaves'. A slave displacement is deleted, when all the stiffness influence coefficients corresponding to it have been added into the system stiffness matrix. The following expression, which is equivalent to (5.92), is used to delete slave displacements:

$$K_{ij}{}^* = K_{ij} - K_{is}K_{sj}/K_{ss} \tag{5.93}$$

where,

$K_{ij}{}^*$ = reduced system stiffness matrix coefficient

s = a slave displacement

N.B. Slave displacements must be eliminated in ascending order.

5.10   COMPUTER PROGRAMS

Computer programs for solving the problems described in the present chapter appear in references [15, 16, and 27].

Using the "PLANESTRESS" computer program of reference [15], the in-plane plate problem of Fig. 6.4, which had a hole in its centre, was analysed on an Apple II microcomputer, and Fig. 5.25 shows the deflected form of this plate.

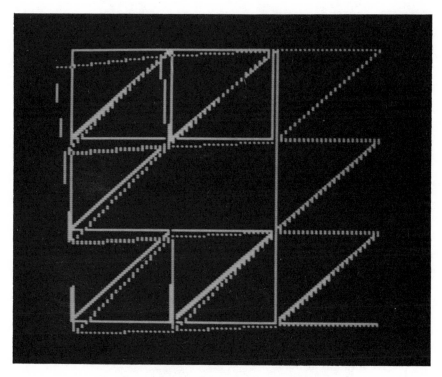

Fig. 5.25 – Deflected form of in-plane plate (Example 6.2).

Other applications of finite elements
The theory described in this chapter can also be applied to many other branches of mathematical physics [13, 19, 41, 42], including:

 a) Heat transfer
 b) Fluid flow
 c) Seepage
 d) Acoustics
 e) Electrostatics, magnetostatics, etc.

EXAMPLES FOR PRACTICE

1. Determine the stiffness matrix for a tapered rod, where the cross-sectional area varies linearly with length by

 (a) Equation (5.16)
 (b) Equation (5.26).

Answer

$$[k] = \frac{(A_1 + A_2)E}{2l} \begin{bmatrix} 1 & -1 \\ -1 & 1 \end{bmatrix}$$

where

A_1 = cross-sectional area at node 1

A_2 = cross-sectional area at node 2.

2. Determine the stiffness matrix for a tapered circular section torque bar, where the diameter varies linearly with length.

Answer

$$[k] = \frac{\pi GC}{10l} \begin{bmatrix} 1 & -1 \\ -1 & 1 \end{bmatrix}$$

where

$C = R_1^4 + R_1^3 R_2 + R_1^2 R_2^2 + R_1 R_2^3 + R_2^4$

G = rigidity modulus

R_1 = radius at node 1

R_2 = radius at node 2.

3. Determine the stiffness matrix for the beam of Fig. 5.14 by using equation (5.16).

It may be assumed that

$$[C^{-1}] = \begin{bmatrix} 1 & 0 & 0 & 0 \\ 0 & -1 & 0 & 0 \\ -3/l^2 & 2/l & 3/l^2 & 1/l \\ 2/l^3 & -1/l^2 & -2/l^3 & -1/l^2 \end{bmatrix}.$$

Answer

$$EI \begin{bmatrix} 12/l^3 & & \text{Symmetrical} & \\ -6/l^2 & 4/l & & \\ -12/l^3 & 6/l^2 & 12/l^3 & \\ -6/l^2 & 2/l & 6/l^2 & 4/l \end{bmatrix}.$$

4. Determine the stiffness matrix for the axisymmetric in-plane annular element of Fig. 5.13 under plane stress, using (equation (5.16).

Answer

$$[k] = \begin{bmatrix} k_{11} & k_{12} \\ \hline k_{21} & k_{22} \end{bmatrix}$$

where

$$k_{11} = CN\{R_2[R_2 \ln(R_2/R_1 - 2(1+v)(R_2 - R_1)]$$
$$+ (1+v)(R_2{}^2 - R_1{}^2)\}$$
$$k_{12} = k_{21} = CN\{-R_1 R_2 \ln(R_2/R_1)\}$$
$$k_{22} = CN\{R_1[R_1 \ln(R_2/R_1) - 2(1+v)(R_2 - R_1)]$$
$$+ (1+v)(R_2{}^2 - R_1{}^2)\}$$
$$CN = \frac{2\pi E t}{(1-v^2)(R_2 - R_1)^2} .$$

5. Determine the $[\tilde{k}]$ and $[C]$ matrices for a uniform thickness axisymmetric plate under flexure, as shown in Fig. 5.26.

   It may be assumed that

   $$w = \alpha_1 + \alpha_2 r + \alpha_3 r^2 + \alpha_4 r^3$$

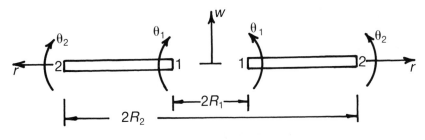

Fig. 5.26.

Answer

$$[C] = \begin{bmatrix} 1 & R_1 & R_1{}^2 & R_1{}^3 \\ 0 & 1 & 2R_1 & 3R_1{}^2 \\ 1 & R_2 & R_2{}^2 & R_2{}^3 \\ 0 & 1 & 2R_2 & 3R_2{}^2 \end{bmatrix} .$$

$$[\tilde{k}] = CN \begin{bmatrix} 0 \\ 0 & \ln(R_2/R_1) \\ 0 & 2(1+v)(R_2-R_1) \\ 0 & 1.5(1+2v)(R_2{}^2-R_1{}^2) \end{bmatrix}$$

$$\begin{matrix} & & \text{Symmetrical} \\ & 4(1+v)(R_2{}^2-R_1{}^2) \\ & 6(1+v)(R_2{}^3-R_1{}^3) & 9(1.25+v)(R_2{}^4-R_1{}^4) \end{matrix}$$

$$CN = \frac{\pi E t^3}{6(1-v^2)}.$$

6. Calculate the reactions and displacements for the plane stress problems of Figs 5.27 and 5.28.

(a) $t = 1 \times 10^{-2}$ m

$E = 2 \times 10^8$ kN/m²

$v = 0.3$.

Answer

$$\{u_F\} = \begin{Bmatrix} u_3{}^o \\ v_3{}^o \\ u_4{}^o \\ v_4{}^o \end{Bmatrix} = \begin{Bmatrix} -0.01349 \\ -0.06175 \\ 0.0260 \\ -0.0427 \end{Bmatrix} \text{cm}$$

$$\{R\} = \begin{Bmatrix} H_{x1} \\ V_{y1} \\ H_{x2} \\ V_{y2} \end{Bmatrix} = \begin{Bmatrix} -450 \\ 250 \\ 250 \\ 0 \end{Bmatrix} \text{kN}$$

where, H_{x1} = Reaction of mode 1 in x^o direction V_{y1} = Reaction of mode 1 in y^o direction.

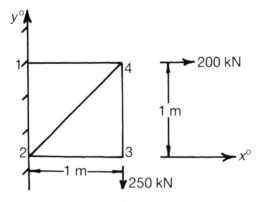

Fig. 5.27.

(b) $t = 2 \times 10^{-2}$ m

$E = 2.08 \times 10^5$ MN/m^2

$v = 0.3$

Answer

$$\{u_F\} = \begin{Bmatrix} u_3^{\,\circ} \\ v_3^{\,\circ} \\ u_4^{\,\circ} \\ v_4^{\,\circ} \end{Bmatrix} = \begin{Bmatrix} 0.0135 \\ -0.0160 \\ 0.0380 \\ -0.0135 \end{Bmatrix} \text{cm}$$

$$\{R\} = \begin{Bmatrix} H_{x1} \\ V_{y1} \\ H_{x2} \\ V_{y2} \end{Bmatrix} = \begin{Bmatrix} -0.833 \\ 0.351 \\ -0.167 \\ -0.351 \end{Bmatrix} \text{MN}$$

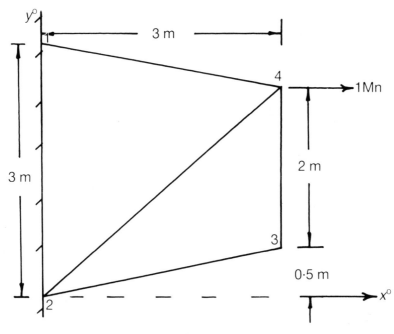

Fig. 5.28.

7. Determine the forces, due to temperature change, in the members of the plane pin-jointed truss of problem 1(a) (Chapter 3), given the following:

(a) There is no 1 MN load at node 4.

(b) $A = 0.001$ m^2, $E = 2 \times 10^{11}$ N/m^2, $\alpha = 12 \times 10^{-6}/°$C.

(c) There is a temperature rise of 50°C in member 1–4, together with a temperature fall of 50°C in member 3–4. (There is no temperature change in member 2–4)

Answer

$$\lfloor u_F \rfloor = \lfloor u_4{}^\circ \quad v_4{}^\circ \rfloor = \lfloor 1.302\text{E-3} \quad 3.668\text{E-5} \rfloor\,\text{m}$$

$$\lfloor F_{1\_4} \quad F_{2\_4} \quad F_{3\_4} \rfloor = \lfloor 6577 \quad -7335 \quad 5370 \rfloor\,\text{N}$$

8. Determine the temperature stresses in the pin-jointed space truss of problem 2(a) of Chapter 3, assuming the following:

    (a) $A = 0.001\ \text{m}^2$, $E = 2 \times 10^{11}\ \text{N/m}^2$, $\alpha = 12 \times 10^{-6}/°\text{C}$

    (b) The 4 MN and 10 MN loads at node 5 are not present.

    (c) Members 1–5 and 4–5 are subjected to a temperature rise of 50°C, and members 2–5 and 3–5 to a temperature fall of 50°C.

Answer

$$\lfloor u_5{}^\circ \quad v_5{}^\circ \quad w_5{}^\circ \rfloor = \lfloor 0 \quad 8.155\text{E-3} \quad 0 \rfloor\,\text{m}$$

$$\lfloor F_{1-5} \quad F_{2-5} \quad F_{3-5} \quad F_{4-5} \rfloor = \lfloor -11260 \quad 15925 \quad 11260$$
$$-15925 \rfloor\,\text{MN}$$

Chapter 6

In-plane quadrilateral elements

This chapter is concerned with the presentation of a four node and an eight node isoparametric quadrilateral element, together with computer analyses of some plane stress and plane strain problems.

The elements which are shown in Fig. 6.1 and 6.2, were introduced because there was a requirement for elements to represent curved boundaries, and also because elements such as the constant stress triangle of Turner et al. [5] were not sufficiently sophisticated to mathematically model stress distributions in areas of steep stress gradient. This was largely because the constant stress triangle was developed on the assumption that the stress and strain were constant in any given direction for any particular element.

This deficiency has been rectified to some extent by the introduction of more sophisticated elements, such as higher order triangles and quadrilaterals, where stress and strain are allowed to vary across the element [7, 8, 43].

In the case of the four node quadrilateral, the displacement variation is linear, and in the case of the eight node quadrilateral it is parabolic. There are, of course, even higher order quadrilaterals, but these will not be described in any detail in the present text.

One of the problems that occur with these elements is that the integrals that appear within the elemental matrices have, in general, got to be determined numerically, and because of this, the computational time required to generate such elements is much larger than that required for the constant stress triangle. Another problem with these elements is that the bandwidth of the system matrices increases in proportion to the number of nodes per element, assuming of course that the total number of elements used to model the plate is the same. However, these deficiencies are relatively minor compared to the advantages gained in the superior stress and strain predictions and also in the improved modelling of the plate, particularly if it has curved boundaries.

Details of these elements were first published by Irons[43]; and in 1968,[7], the elements we are concerned with in the present chapter were published.

Elements such as these have been used to successfully model problems varying from earth dams to supertankers (see section 5.2).

6.1 AN ISOPARAMETRIC ELEMENT

An *isoparametric element* is said to be one whose displacement functions and geometry can both be described by the same matrix of shape functions. For example, for a two-dimensional element,

$$\{u_{(x,y)}\} = \begin{Bmatrix} u_i{}^o \\ v_i{}^o \end{Bmatrix} = [N]\{u_i{}^o\} \tag{6.1}$$

and

$$\begin{Bmatrix} x^o \\ y^o \end{Bmatrix} = [N]\begin{Bmatrix} x_i{}^o \\ y_i{}^o \end{Bmatrix} \tag{6.2}$$

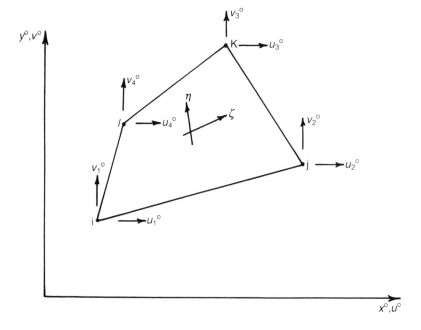

Fig. 6.1 – Four node quadrilateral element.

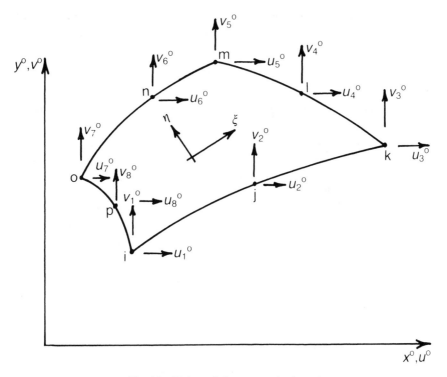

Fig. 6.2 – Eight node isoparametric element.

where

[N] = a matrix of shape functions, which is usually expressed in normalised coordinates

$$\{u_i^o\}^T = \lfloor u_1^o \quad v_1^o \quad u_2^o \ldots u_{NN}^o \quad v_{NN}^o \rfloor$$

= a vector of nodal displacements in global coordinates.

$$\begin{Bmatrix} x_i^o \\ y_i^o \end{Bmatrix} = \lfloor x_1^o \quad y_1^o \quad x_2^o \quad y_2^o \ldots \ldots x_{NN}^o \quad y_{NN}^o \rfloor$$

= a vector of nodal coordinates,

NN = number of nodes.

Already, we have met an isoparametric element, namely, the rod element of Chapters 3 and 5, whose displacement and geometry can both be expressed by the same matrix of displacement functions, as follows:

$$\{u_{(x,y)}\} = \{u\} = \lfloor (1-\xi) \quad \xi \rfloor \begin{Bmatrix} u_1 \\ u_2 \end{Bmatrix} \tag{6.3}$$

and

$$\{x\} = \lfloor (1-\xi) \quad \xi \rfloor \begin{Bmatrix} x_1 \\ x_2 \end{Bmatrix}. \tag{6.4}$$

From (6.3) and (6.4) it can be seen that at $\xi = 0$, $x = x_1$ and $u = u_1$, and at $\xi = 1$, $x = x_2$ and $u = u_2$, which are as required. That is, the plane rod of Chapters 3 and 5 is said to be isoparametric, as its displacement and geometry are both related by the same matrix of shape functions and the same number of nodes.

Similarly, if more nodes are used to define the geometry of the element than are used to define its displacement distribution, the element is said to be *super-parametric* and vice versa for a *sub-parametric* element.

The isoparametric approach in finite element methods opens up a much more sophisticated family of elements, as they lend themselves to analysing much more complex shapes. This has been further enhanced by both Zienkiewicz[13] and Irons & Ahmad [19], who have produced a large selection of shape functions, to make it possible to develop many and various elements.

6.2 FOUR NODE QUADRILATERAL ELEMENT

The assumed displacement functions are

$$u^o = \alpha_1 + \alpha_2 x^o + \alpha_3 y^o + \alpha_4 x^o y^o$$
$$v^o = \alpha_5 + \alpha_6 x^o + \alpha_7 y^o + \alpha_8 x^o y^o,$$

and Ergatoudis *et al.*[7] have shown that these displacement functions can be expressed by the following equations:

$$\left\{ \begin{array}{c} u^o \\ v^o \end{array} \right\} = \left[\begin{array}{cccccccc} N_1 & 0 & N_2 & 0 & N_3 & 0 & N_4 & 0 \\ 0 & N_1 & 0 & N_2 & 0 & N_3 & 0 & N_4 \end{array} \right] \left\{ \begin{array}{c} u_1{}^o \\ v_1{}^o \\ u_2{}^o \\ v_2{}^o \\ u_3{}^o \\ v_3{}^o \\ u_4{}^o \\ v_4{}^o \end{array} \right\} \tag{6.5}$$

$$= [N]\{u_i{}^o\}$$

Similarly for $\lfloor x^o \quad y^o \rfloor^{\mathrm{T}}$,

where

$[N]$ = a matrix of shape functions $\tag{6.6}$

$\{u_1{}^o\}$ = a vector of nodal displacements.

The nodal displacements are u^o and v^o, and these are in the x^o and y^o directions, respectively. The coordinates ξ and η point from i to j and j to k respectively, and the values for ξ and η at the nodes are as follows:

Node 1 – $\xi = -1$ and $\eta = -1$
Node 2 – $\xi = 1$ and $\eta = -1$
Node 3 – $\xi = 1$ and $\eta = 1$
Node 4 – $\xi = -1$ and $\eta = 1.$

The elements of the matrix of shape functions $[N]$ are given by

$$N_1 = \tfrac{1}{4}(1-\xi)(1-\eta)$$
$$N_2 = \tfrac{1}{4}(1+\xi)(1-\eta)$$
$$N_3 = \tfrac{1}{4}(1+\xi)(1+\eta)$$
$$N_4 = \tfrac{1}{4}(1-\xi)(1+\eta).$$

The stiffness matrix is given by

$$[k^\circ] = \int\int [B]^{\mathrm{T}}[D][B]\,\mathrm{d}x\,\mathrm{d}y$$

$$= \int_{-1}^{1}\int_{-1}^{1} [B]^{\mathrm{T}}[D][B]\,\det|J|\,\mathrm{d}\xi\,\mathrm{d}\eta, \tag{6.7}$$

where

$$[B] = [B_1\ B_2\ B_3\ B_4] \tag{6.8}$$

$$[B_i] = \begin{bmatrix} \dfrac{\partial N_i}{\partial x^\circ} & 0 \\[2mm] 0 & \dfrac{\partial N_i}{\partial y^\circ} \\[2mm] \dfrac{\partial N_i}{\partial y^\circ} & \dfrac{\partial N_i}{\partial x^\circ} \end{bmatrix} \tag{6.9}$$

$$\begin{Bmatrix} \dfrac{\partial N_i}{\partial x^\circ} \\[2mm] \dfrac{\partial N_i}{\partial y^\circ} \end{Bmatrix} = [J^{-1}] \begin{Bmatrix} \dfrac{\partial N_i}{\partial \xi} \\[2mm] \dfrac{\partial N_i}{\partial \eta} \end{Bmatrix} \tag{6.10}$$

$$\det|J| = \frac{\partial(x^\circ, y^\circ)}{\partial(\xi, \eta)} \begin{vmatrix} \dfrac{\partial x^\circ}{\partial \xi} & \dfrac{\partial y^\circ}{\partial \xi} \\[2mm] \dfrac{\partial x^\circ}{\partial \eta} & \dfrac{\partial y^\circ}{\partial \eta} \end{vmatrix} = \text{a Jacobian determinant } [44]$$

that is

$$[J] = \begin{bmatrix} \dfrac{\partial N_1}{\partial \xi} & \dfrac{\partial N_2}{\partial \xi} & \dfrac{\partial N_3}{\partial \xi} & \dfrac{\partial N_4}{\partial \xi} \\[2ex] \dfrac{\partial N_1}{\partial \eta} & \dfrac{\partial N_2}{\partial \eta} & \dfrac{\partial N_3}{\partial \eta} & \dfrac{\partial N_4}{\partial \eta} \end{bmatrix} \begin{bmatrix} x_1^{\,\circ} & y_1^{\,\circ} \\ x_2^{\,\circ} & y_2^{\,\circ} \\ x_3^{\,\circ} & y_3^{\,\circ} \\ x_4^{\,\circ} & y_4^{\,\circ} \end{bmatrix}$$

$x_i^{\,\circ}$ and $y_i^{\,\circ}$ are the coordinates of the ith node.

$$\frac{\partial N_1}{\partial \xi} = -\tfrac{1}{4}(1-\eta)$$

$$\frac{\partial N_2}{\partial \xi} = \tfrac{1}{4}(1-\eta)$$

$$\frac{\partial N_3}{\partial \xi} = \tfrac{1}{4}(1+\eta)$$

$$\frac{\partial N_4}{\partial \xi} = -\tfrac{1}{4}(1+\eta) \tag{6.11}$$

$$\frac{\partial N_1}{\partial \eta} = -\tfrac{1}{4}(1-\xi)$$

$$\frac{\partial N_2}{\partial \eta} = -\tfrac{1}{4}(1+\xi)$$

$$\frac{\partial N_3}{\partial \eta} = \tfrac{1}{4}(1+\xi)$$

$$\frac{\partial N_4}{\partial \eta} = \tfrac{1}{4}(1-\xi)$$

$$[D] = E^1 \begin{bmatrix} 1 & \mu & 0 \\ \mu & 1 & 0 \\ 0 & 0 & \gamma \end{bmatrix} \tag{6.12}$$

where the symbols have the same meanings as in section 5.2.

By substitution of (6.8) to (6.12) into (6.7), the stiffness matrix [k] can be determined. Integration will have to be carried out numerically, and one of the most suitable methods of integration is through *Gauss-Legendre* [13, 19, 34, 41].

Numerical integration through Gauss-Legendre, depends on the Gauss points used and the order of the equation to be integrated, and further work on this is done in references[13, 19, 34, 41].

In the program presented in this chapter, it was found sufficient to use two Gauss points in the ξ direction and two Gauss points in the η direction, making a total of four Gauss points per element.

6.3   EIGHT NODE ISOPARAMETRIC QUADRILATERAL

This is a much more sophisticated element than the four node quadrilateral as it allows for a parabolic variation in displacement. The element has four of its nodes at its corners and the other four nodes at its 'mid-sides', as shown in Fig. 6.2. The 'mid-side' nodes need not be exactly at the mid-sides of the element. Each node has two degrees of freedom, making a total of sixteen degrees of freedom per element.

The assumed displacement functions are

$$u^\circ = \alpha_1 + \alpha_2 x + \alpha_3 y + \alpha_4 xy + \alpha_5 x^2 + \alpha_6 y^2 + \alpha_7 x^2 y + \alpha_8 y^2 x$$

$$v^\circ = \alpha_9 + \alpha_{10} x + \alpha_{11} y + \alpha_{12} xy + \alpha_{13} x^2 + \alpha_{14} y^2 + \alpha_{15} x^2 y + \alpha_{16} y^2 x$$

which can be put in the following matrix form:

$$\left\{ \begin{matrix} u^\circ \\ v^\circ \end{matrix} \right\} = \begin{bmatrix} N_1 & 0 & N_2 & 0 & N_3 & 0 & N_4 & \ldots & N_8 & 0 \\ 0 & N_1 & 0 & N_2 & 0 & N_3 & 0 & \ldots & 0 & N_8 \end{bmatrix} \left\{ \begin{matrix} u_1^\circ \\ v_1^\circ \\ u_2^\circ \\ v_2^\circ \\ u_3^\circ \\ v_3^\circ \\ \vdots \\ u_8^\circ \\ v_8^\circ \end{matrix} \right\} \tag{6.13}$$

$$= [N]\{u_i^\circ\},$$

where

$[N]$ = a matrix of shape functions

$\{u_i^\circ\}$ = a vector of nodal displalacements

$N_1 = -\frac{1}{4}(1-\xi)(1-\eta)(\xi+\eta+1)$

$N_2 = \frac{1}{2}(1-\xi^2)(1-\eta)$

$N_3 = \frac{1}{4}(1+\xi)(1-\eta)(\xi-\eta-1)$

$N_4 = \frac{1}{2}(1-\eta^2)(1+\xi)$

$N_5 = \frac{1}{4}(1+\xi)(1+\eta)(\xi+\eta-1)$

$N_6 = \frac{1}{2}(1-\xi^2)(1+\eta)$

$N_7 = -\frac{1}{4}(1-\xi)(1+\eta)(\xi-\eta+1)$

$N_8 = \frac{1}{2}(1-\eta^2)(1-\xi)$.

The curvilinear coordinates ξ and η point from p to l and j to n respectively, and the values of ξ and η at the nodes are

Node i   *Node j*

$\xi = -1 \quad \eta = -1 \qquad \xi = 0 \qquad \eta = -1$

Node k		*Node n*	
$\xi = 1$	$\eta = -1$	$\xi = 0$	$\eta = 1$

Node l		*Node o*	
$\xi = 1$	$\eta = 0$	$\xi = -1$	$\eta = 1$

Node m		*Node p*	
$\xi = 1$	$\eta = 1$	$\xi = -1$	$\eta = 0.$

The stiffness matrix is obtained from (6.7), where

$$[B] = [B_1 \quad B_2 \quad B_3 \quad B_4 \quad B_5 \quad B_6 \quad B_7 \quad B_8] \tag{6.14}$$

$[B_i]$ is as in equation (6.9)

and $\dfrac{\partial N_i}{\partial x^0}$ and $\dfrac{\partial N_i}{\partial y^0}$ are as in equation (6.10), where

$$[J] = \begin{vmatrix} \dfrac{\partial N_1}{\partial \xi} & \dfrac{\partial N_2}{\partial \xi} & \dfrac{\partial N_3}{\partial \xi} & \cdots & \dfrac{\partial N_8}{\partial \xi} \\[2mm] \dfrac{\partial N_1}{\partial \eta} & \dfrac{\partial N_2}{\partial \eta} & \dfrac{\partial N_3}{\partial \eta} & \cdots & \dfrac{\partial N_8}{\partial \eta} \end{vmatrix} \begin{bmatrix} x_1^0 & y_1^0 \\ x_2^0 & y_2^0 \\ x_3^0 & y_3^0 \\ \vdots & \vdots \\ x_8^0 & y_8^0 \end{bmatrix} \tag{6.15}$$

$$\frac{\partial N_1}{\partial \xi} = 0.25(1 - \eta)(2\xi + \eta)$$

$$\frac{\partial N_1}{\partial \eta} = 0.25(1 - \xi)(\xi + 2\eta)$$

$$\frac{\partial N_2}{\partial \xi} = -\xi(1 - \eta) \tag{6.16}$$

$$\frac{\partial N_2}{\partial \eta} = -0.5(1 - \xi^2).$$

Similarly for the other derivatives.

The matrix $[D]$ is given in section 5.2 and use of this matrix, together with the other relevant matrices, gives an expression for the stiffness matrix $[k^0]$.

Integration was carried out using three Gauss points in the ξ direction and three in the η direction, making a total of nine Gauss points per element.

6.4 COMPUTER PROGRAMS

Two computer programs "stress 4" and "stress 8", are used for analysing a number of in-plane plate problems through the use of isoparametric elements, and comparisons are made with the results obtained from these two elements, together with those obtained from the constant stress triangle.

The computer programs calculate nodal displacements, together with direct and shear stresses at a number of points in each element. The direct stresses are determined in the x° and y° directions, and the shear stresses in the x°–y° planes.

6.5 FOUR NODE QUADRILATERAL SOLUTION

This program uses an element which allows for a linear variation in displacement. The element has four nodes with two degrees of freedom per node, making a total of eight degrees of freedom per element. Thus, if the number of nodes is NJ, there are $2*NJ$ degrees of freedom.

In determining the system stiffness matrix, advantage was taken of any sparsity in the matrix by storing only the upper half of the matrix that was contained within the band, as described in Chapter 1. Thus, by choosing a suitable system of nodal numbering, a more efficient computer solution was realised. For example, for the plate of Fig. 6.5, by choosing a system of nodal numbering which numerically increased across the smaller dimension, in the manner shown, the bandwidth was much smaller than that which would have occurred had the system of nodal numbering increased numerically along the longer dimension. This is because the bandwidth depends on the largest variation in nodal numbers for any particular element.

6.5.1 Element details

The nodal points defining each element were fed in a counter clockwise direction, as follows:

   i   j   k   l

where

   ξ points from i to j and l to k

and

   η points from i to l and j to k,

as shown in Fig. 6.1.

The following notation is used in the results for stresses:

   σ_x = direct stress in x° direction

   σ_y = direct stress in y° direction

   τ_{xy} = shear stress in the x°–y° plane.

To demonstrate the capabilities of the element, Examples 6.1 to 6.3 will be considered.

6.5.2   Example 6.1

Determine the nodal displacements and stresses for the plate shown in Fig. 6.3, fixed at nodes 1 and 2, where one element is adopted.

$$E = 6.93 \times 10^4 \text{ MN/m}^2 \qquad v = 0.3 \qquad t = 2 \times 10^{-2} \text{ m}$$

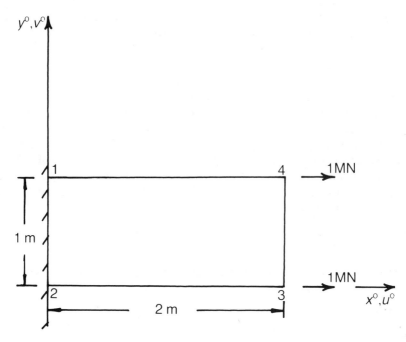

Fig. 6.3 – One element plate.

Element details (nodal numbering)

$$i = 1 \qquad j = 2 \qquad k = 3 \qquad l = 4$$

(fed in a counter-clockwise direction). This choice of nodal numbering will make ξ point in direction from 1 to 2 and η from 2 to 3.

N.B: If the nodal numbering were 2, 3, 4, 1, then ξ would have been in direction from 2 to 3 and η from 3 to 4.

Results
(a) *Plane stress*

	Nodal displacements (m)	
Node i	$u_i{}^o$	$v_i{}^o$
1	0	0
2	0	0
3	2.812E-3	3.096E-4
4	2.812E-3	−3.096E-4

Stresses (MN/m^2)

Element	ξ	η	σ_x	σ_y	τ_{xy}
1	−1	−1	107.1	32.1	−4.13
1	−1	0	100	8.55	−4.13
1	−1	1	92.1	−15.0	−4.13
1	0	−1	107.1	32.1	0
1	0	0	100	8.55	0

(b) *Plane strain*

	Nodal displacements (m)	
Node	u^o	v^o
1	0	0
2	0	0
3	2.47E-3	3.84E-4
4	2.47E-3	−3.84E-4

Stresses (MN/m^2)

Element	ξ	η	σ_x	σ_y	τ_{xy}
1	−1	−1	115.4	49.4	−10.0
	−1	0	100	13.6	−10.0
	−1	1	84.6	−22.2	−10.0
	0	−1	115.4	49.4	0
1	0	0	100	13.6	0

These results compare favourably with those of the simplex triangular element of reference [15], except that they appear to be more precise. For example, for plane stress, σ_x appears to be about $\pm 7\%$ and the u^o displacements are underestimated by 2.6%. This is very encouraging, especially as only one element was used.

6.5.3   Example 6.2

Determine the nodal displacements and stresses for the square plate with a hole in its centre, under plane stress, as shown in Fig. 6.4. The plate is firmly fixed at nodes 13, 14, 15, and 16, and the following also apply:

$$E = 1 \times \times 10^{11}/\text{N/m}^2$$
$$v = 0.3$$
$$t = 2 \times 10^{-2} \text{ m}$$

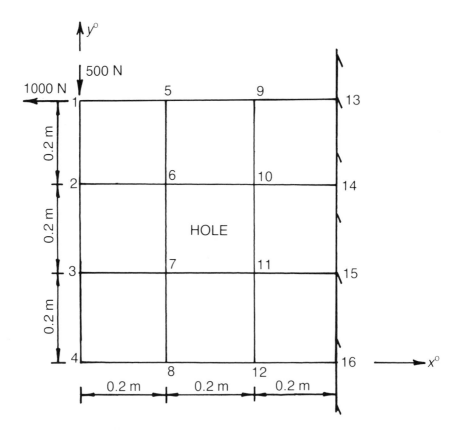

Fig. 6.4 – Plate with hole.

Element details (Nodal numbering)

i	j	k	l
1	2	6	5
2	3	7	6
3	4	8	7
5	6	10	9
7	8	12	11
9	10	14	13
10	11	15	14
11	12	16	15

Results

Node i	Nodal Displacements (m)	
	$u_i^{\,o}$	$v_i^{\,o}$
1	$-3.68\text{E-}6$	$-4.82\text{E-}6$
2	$-6.74\text{E-}7$	$-3.81\text{E-}6$
3	$1.90\text{E-}7$	$-3.33\text{E-}6$
4	$1.19\text{E-}6$	$-3.32\text{E-}6$
5	$-2.32\text{E-}6$	$-2.09\text{E-}6$
6	$-5.69\text{E-}7$	$-2.08\text{E-}6$
7	$9.67\text{E-}8$	$-2.19\text{E-}6$
8	$1.18\text{E-}6$	$-2.11\text{E-}6$
9	$-1.23\text{E-}6$	$-7.49\text{E-}7$
10	$-1.92\text{E-}7$	$-4.21\text{E-}7$
11	$-4.45\text{E-}8$	$-5.12\text{E-}7$
12	$8.10\text{E-}7$	$-7.49\text{E-}7$
13	0	0
14	0	0
15	0	0
16	0	0

Some of the stresses (N/m^2) are as follows:

Element	ξ	η	σ_x	σ_y	τ_{xy}
1	-1	-1	576359	-320105	-51407
1	-1	0	665557	-41361	67239
1	-1	1	754755	237383	185885
1	0	-1	227399	-431772	-146180

Element	ξ	η	σ_x	σ_y	τ_{xy}
8	0	1	−213349	−68272	119482
8	1	−1	−409178	−12329	−19971
8	1	0	−430320	−78399	60987
8	1	1	−451462	−144468	141945

These results appear to be quite different to the simplex triangular element of reference [15].

6.5.4 Example 6.3

Determine the nodal displacements and stresses for the plane stress problem of Fig. 6.5. The elements have been distorted to demonstrate their distorting capabilities, but caution must be exercised when carrying out this feature.

$$E = 2.07 \times 10^{11} \qquad v = 0.3 \qquad t = 0.254 \text{ cm}$$

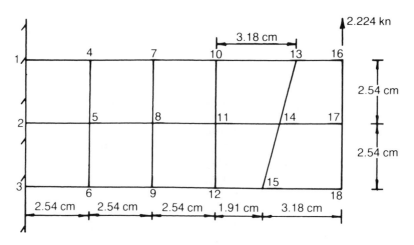

Fig. 6.5 – Cantilever plate.

Element details (nodal numbering)

i	j	k	l
1	2	5	4
2	3	6	5
4	5	8	7
5	6	9	8
7	8	11	10
8	9	12	11
10	11	14	13
11	12	15	14
13	14	17	16
14	15	18	17

Results

Node	Displacements (mm)	
	u^o	v^o
1	0	0
2	0	0
3	0	0
4	−0.0248	0.0200
5	−2.16E-6	0.0158
6	0.0248	0.0200
7	−0.0447	0.0592
8	2.00E-5	0.0569
9	0.0447	0.0594
10	−0.0589	0.1163
11	1.42E-4	0.1146
12	0.0587	0.1163
13	−0.0683	0.2022
14	2.77E-5	0.1836
15	0.0658	0.1669
16	−0.0701	0.2591
17	−8.71E-5	0.2565
18	0.0701	0.2591

Some of the stress values are as follows:

Element	ξ	η	σ_x	σ_y	τ_{xy}
			Stresses (MN/m^2)		
1	−1	−1	−221.7	−66.5	62.6
1	−1	0	−216.1	−47.9	23.8
1	−1	1	−210.5	−29.2	−15.0
1	0	−1	−110.9	−33.3	56.0
10	0	1	14.5	−2.0	10.2
10	1	−1	27.1	−2.6	20.5
10	1	0	27.8	−0.2	15.0
10	1	1	28.6	2.3	9.5

According to elementary theory,

$$\sigma_x(\text{max}^m) = +258 \text{ MN/m}^2$$

$$\sigma_x(\text{min}^m) = -258.6 \text{ MN/m}^2$$

$$v^o{}_{16} = u^o{}_{17} = v^o{}_{18} = 0.264 \text{ mm}.$$

These seem to compare favourably with the results from the computer program, and are considerably better than the computer results for the simplex triangular element of reference [15].

6.6　EIGHT NODE QUADRILATERAL SOLUTION

This program adopts an element with eight nodes and two degrees of freedom per node, making a total of 16 degrees of freedom per element. Thus, if the number of nodes is NJ, there are $2*NJ$ degrees of freedom.

In determining the system stiffness matrix, advantage was taken of any sparsity in the matrix by storing only the upper half of the matrix that was contained within the band, as described in references [15] and [16]. To take full advantage of this feature, the preferred method of nodal numbering for a long plate is across the plate as shown in Figs 6.5 and 6.8, rather than along the length. The reason for this is that the half bandwidth is dependent on the largest variation in node numbers for any particular element.

The direct and shear stresses are calculated at several points in an element, and the symbols below have the following meanings:-

σ_x = direct stress in the x^o direction

σ_y = direct stress in the y^o direction

τ_{xy} = shear stress in the $x^o - y^o$ plane.

6.6.1　Element details

The nodal points defining each element were fed in a counter-clockwise direction, as follows:

　　　i　　j　　k　　l　　m　　n　　o　　p

where

　　　ξ points from p to l and η points from j to n, as shown in Fig. 6.2. To demonstrate the capabilities of the element, Examples 6.4 to 6.6 will be considered.

6.6.2　Example 6.4

Determine the nodal displacements and stresses for the problem of Example 6.1.

Element details (nodal numbering)

　　　i = 1　　j = 2　　k = 3　　l = 4　　m = 5　　n = 6
　　　o = 7　　p = 8

(fed in a counter-clockwise direction, starting from a CORNER NODE). This

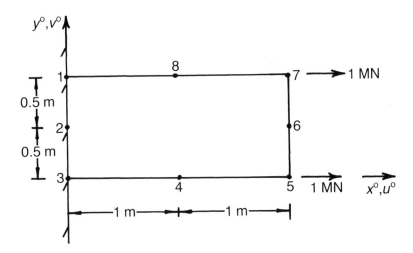

Fig. 6.6 – One element plate.

choice of nodal numbering will cause ξ to point in direction from 8 to 4 and η from 2 to 6.

Note
If the nodal numbering were 3, 4, 5, 6, 7, 8, 1, 2, 3, then ξ would have pointed from 2 to 6 and η from 4 to 8.

Results

(a) *Plane stress*

Node i	Nodal displacements (m)	
	$u_i{}^o$	$u_i{}^o$
1	0	0
2	0	0
3	0	0
4	1.45E-3	2.16E-4
5	2.97E-3	2.43E-4
6	2.78E-3	−2.58E-11
7	2.97E-3	−2.43E-4
8	1.45E-3	−2.16E-4

Stresses (MN/m^2)

Element	ξ	η	σ_x	σ_y	τ_{xy}
1	-1	-1	107.7	32.3	-8.27
1	-1	0	103.4	1.11	6.94
1	-1	1	107.7	-1.34	22.2
1	0	-1	100.5	30.1	0
1	0	0	96.1	-1.1	0
1	0	1	100.5	-3.5	0
1	1	-1	107.7	32.3	8.27
1	1	0	103.4	1.11	6.94
1	1	1	107.7	-1.34	-22.2

(b) *Plane strain*

Node	Nodal displacements (m)	
i	$u_i{}^o$	$v_i{}^o$
1	0	0
2	0	0
3	0	0
4	1.25E-3	2.75E-4
5	2.62E-3	3.07E-4
6	2.51E-3	-1.76E-11
7	2.62E-3	-3.07E-4
8	1.25E-3	-2.75E-4

Stress (MN/M^2)

Element	ξ	η	σ_x	σ_y	τ_{xy}
1	-1	-1	110.0	47.2	-20.7
1	-1	0	100.3	1.1	3.9
1	-1	1	110.0	0.4	28.6
1	0	-1	104.7	44.9	0
1	0	0	95.0	-1.2	0
1	0	1	104.7	-1.9	0
1	1	-1	110.0	47.2	20.7
1	1	0	100.3	1.1	-3.9
1	1	1	110.0	0.4	-28.6

These stresses and displacements compare favourably with the 4 node quadrilateral element and also the 3 node simplex triangular element.

6.6.3   Example 6.5

Determine the nodal displacements and stresses for the problem of Example 6.2

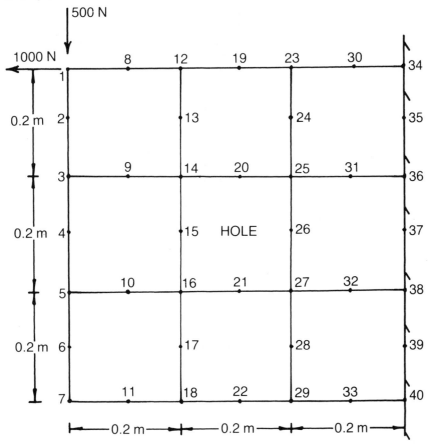

Fig. 6.7 – Square plate with square hole.

Element details (nodal numbering)
1, 2, 3, 9, 14, 13, 12, 8
3, 4, 5, 10, 16, 15, 14, 9
5, 6, 7, 11, 18, 17, 16, 10
12, 13, 14, 20, 25, 24, 23, 19
16, 17, 18, 22, 29, 28, 27, 21
23, 24, 25, 31, 36, 35, 34, 30
25, 26, 27, 32, 38, 37, 36, 31
27, 28, 29, 33, 40, 39, 38, 32

Results

Node	Nodal displacements (m)	
i	$u_i{}^\circ$	$v_i{}^\circ$
1	− 5.23E-6	− 6.66E-6
2	− 1.80E-6	− 5.40E-6
3	− 6.05E-7	− 4.36E-6
4	− 1.16E-7	− 3.89E-6
5	2.22E-7	− 3.77E-6
6	6.92E-7	− 3.77E-6
7	1.27E-6	− 3.77E-6
8	− 3.72E-6	− 3.57E-6
9	− 5.00E-7	− 3.44E-6
10	1.90E-7	− 3.22E-6
11	1.28E-6	− 3.20E-6
12	− 2.52E-6	− 2.34E-6
13	− 1.52E-6	− 2.28E-6
14	− 4.02E-7	− 2.38E-6
15	− 4.21E-8	− 2.63E-6
16	6.58E-8	− 2.60E-6
17	6.40E-7	− 2.53E-6
18	1.31E-6	− 2.5E-6
19	− 1.95E-6	− 1.49E-6
20	− 3.13E-7	− 1.23E-6
21	− 1.20E-7	− 1.55E-6
22	1.26E-6	− 1.64E-6
23	− 1.30E-6	− 7.63E-7
24	− 7.42E-7	− 5.72E-7
25	− 2.36E-7	− 4.72E-7
26	− 6.46E-8	− 4.67E-7
27	− 6.39E-8	− 5.76E-7
28	3.79E-7	− 6.82E-7
29	9.20E-7	− 8.33E-7
30	− 6.32E-7	− 3.10E-7
31	− 1.37E-7	− 1.50E-7
32	− 2.91E-8	− 1.94E-7
33	4.66E-7	− 2.92E-7
34	0	0
35	0	0
36	0	0
37	0	0
38	0	0
39	0	0
40	0	0

Stresses (MN/m^2)

Element	ξ	η	σ_x	σ_y	τ_{xy}
1	-1	-1	1.37E6	-919334	-203896
1	-1	0	1.42E6	205149	-195243
1	-1	1	1.11E6	220434	-244485
1	0	-1	-75496	-1.17E6	40281
8	0	1	-221114	-70756	46607
8	1	-1	-437299	32146	28425
8	1	0	-487536	-85652	45156
8	1	1	-526593	-168510	63195

The results for this plate show that the mesh for the 4 node quadrilateral analysis of Example 6.2 was unsatisfactory and that of the simplex triangular element of reference {15} was totally inadequate. Even for the present case, a much more refined mesh was preferable.

6.6.4 Example 6.6

Determine the nodal displacements and stresses for the cantilever plate of Example 6.3. The isoparametric elements of Fig. 6.2 have deliberately been distorted to test their distorting capabilities.

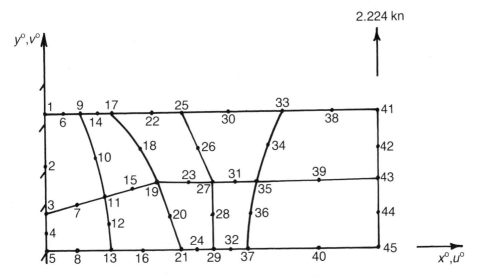

Fig. 6.8 – Isoparametric mesh for cantilever.

Element details (nodal numbering)
1, 2, 3, 7, 11, 10, 9, 6
3, 4, 5, 8, 13, 12, 11, 7
9, 10, 11, 15, 19, 18, 17, 14

11, 12, 13, 16, 21, 20, 19, 15
17, 18, 19, 23, 27, 26, 25, 22
19, 20, 21, 24, 29, 28, 27, 23
25, 26, 27, 31, 35, 34, 33, 30
27, 28, 29, 32, 37, 36, 35, 31
33, 34, 35, 39, 43, 42, 41, 38
35, 36, 37, 40, 45, 44, 43, 39

Results

Node	Nodal displacements (mm)	
i	u_i°	v_i°
1	0	0
2	0	0
3	0	0
4	0	0
5	0	0
6	-7.54E-3	4.11E-3
7	4.57E-3	5.64E-3
8	0.0155	9.37E-3
9	-0.0152	9.40E-3
10	-9.58E-3	0.0109
11	5.82E-3	0.0165
12	0.0156	0.0189
13	0.0287	0.0230
14	-0.0230	0.0151
15	3.15E-3	0.0307
16	0.0406	0.0432
17	-0.0290	0.0229
18	-0.0183	0.0361
19	2.25E-4	0.047
20	0.0231	0.0582
21	0.0508	0.0683
22	-0.0406	0.0432
23	1.07E-4	0.0714
24	0.0554	0.0831
25	-0.0508	0.0683
26	-0.0267	0.0810
27	-1.05E-4	0.0960
28	0.0290	0.0968
29	0.0594	0.0986
30	-0.0630	0.1153

Node	Nodal displacements (m)	
i	$u_i^{\,o}$	$v_i^{\,o}$
31	9.63E-5	0.1201
32	0.0632	0.1151
33	−0.0721	0.1697
34	−0.0338	0.1496
35	2.16E-5	0.1415
36	0.0328	0.1346
37	0.0665	0.1326
38	−0.0777	0.2293
39	−5.31E-5	0.2083
40	0.0762	0.2090
41	−0.0798	0.2946
42	−0.0389	0.2921
43	9.09E-5	0.2921
44	0.0386	0.2921
45	0.0803	0.2946

Stresses (MN/m^2)

Element	ζ	η	σ_x	σ_y	τ_{xy}
1	−1	−1	−267.7	−80.3	44.3
1	−1	0	−254.2	−21.2	30.5
1	−1	1	−238.5	45.4	15.4
1	0	−1	−68.3	−20.5	17.3
10	0	1	−2.3	−18.5	21.8
10	1	−1	98.7	−2.1	8.4
10	1	0	56.1	2.0	2.6
10	1	1	3.9	−25.8	10.3

This problem shows the two quadrilateral elements to be vastly superior to the simplex triangular element, and in particular the eight node quadrilateral element. The problem also shows that the elements can be distorted to some extent. Care, however, [13] must be exercised in not distorting the elements too much, as numerical errors can result.

6.7 OTHER HIGHER ORDER ELEMENTS

In addition to the elements described in the present chapter, there are a number of other higher order elements. Some of these include the cubic quadrilateral of Fig. 6.9 and the parabolic and cubic triangles of Figs 6.10 and 6.11.

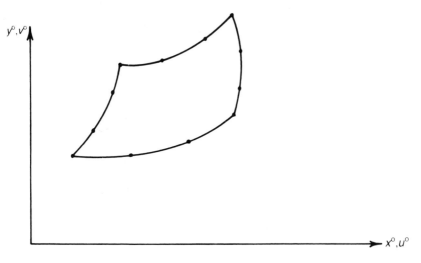

Fig. 6.9 – Cubic quadrilateral.

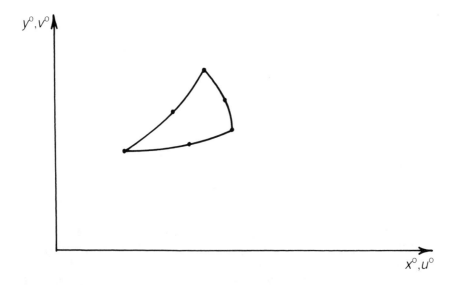

Fig. 6.10 – Parabolic triangle.

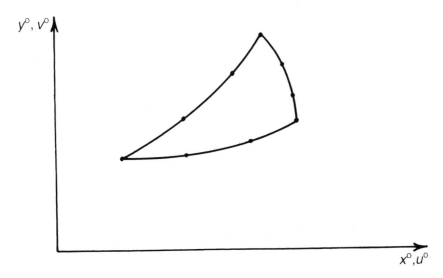

Fig. 6.11 – Cubic triangle.

The advantages of these elements over their simplex equivalents is that they are capable of superior stress predictions and also that they can represent curved boundaries.

Disadvantages of these elements include the longer computational time required to generate them, and the increased bandwidth of the system matrices for plates modelled by the same number of elements.

Chapter 7

Vibrations of structures

The method considered here is an important branch of the finite element analysis of structures, and, like static analysis, it is an extension of the matrix displacement method.

In addition to needing the stiffness matrices, it will now be necessary to introduce the mass matrix as well, and like the stiffness matrix, the elemental mass matrix will be derived by energy methods.

The *elemental mass matrix*, which is always symmetrical, is a matrix of equivalent nodal masses that dynamically represent the actual distributed mass of the element. For example, the mass of a one-dimensional uniform rod of cross-sectional area A and length l is ρAl, and this can be shown to be equivalent to the following elemental mass matrix.

$$
\{m\} = \rho Al \begin{array}{cc} u_1 & u_2 \\ \begin{bmatrix} \frac{1}{3} & \frac{1}{6} \\ \frac{1}{6} & \frac{1}{3} \end{bmatrix} & \begin{array}{c} u_1 \\ u_2 \end{array} \end{array}
$$

where, $\rho = $ density of the element material. This matrix is of order 2×2, because there are two nodal displacements for the element, and it should be noted that the sum of its terms add up to ρAl. Thus, the mass which in fact is uniformly distributed over the length of the element, is represented dynamically by the equivalent mass matrix above.

An alternative approximation to the elemental mass matrix is the *lumped mass matrix*. The usual process adopted in deriving the latter is to distribute the mass of the element evenly between its nodal points in the form of concentrated or lumped masses. For example, the lumped mass matrix for the plane rod is given by

$$[m] = \rho Al \begin{bmatrix} \frac{1}{2} & 0 \\ 0 & \frac{1}{2} \end{bmatrix} \begin{matrix} u_1 \\ u_2 \end{matrix} .$$

This can be seen to have quite a different value from the elemental mass matrix, but it does have computational advantages because of its diagonal form. It is particularly useful when a structure of small mass supports several large concentrated masses, and the influence of the former is small compared with the latter, so far as the effects of mass are concerned. Further work in this chapter will be concerned with the elemental mass matrix and its applications to vibration problems.

Transformation of the elemental mass matrix from local to global coordinates and assemblage of the mass matrix of the entire structure, are carried out in a manner similar to that adopted for the stiffness matrices. That is,

$$[m^\circ] = [\Xi]^T [m] [\Xi]$$

and

$$[M^\circ] = \sum [m^\circ]$$

where

$[m]$ = elemental mass matrix

$[M^\circ]$ = mass matrix for the entire structure in global coordinates.

7.1 EQUATIONS OF MOTION

From Newton's 2nd law of motion,

Force = mass × acceleration, which when applied to a structure in matrix notation becomes

$$\{P_{(t)}\} - [k]\{u_i\} - [C_v]\{\dot{u}_i\} = [m]\{\ddot{u}_i\} \tag{7.1}$$

where

$\{P_{(t)}\}$ = a vector of forcing functions

$\{\dot{u}_i\}$ = a vector of nodal velocities

$\{\ddot{u}_i\}$ = a vector nodal accelerations

$[C_v]$ = a matrix containing viscous damping terms.

If there is no damping, as is assumed for all problems in this chapter, then (7.1) becomes

$$\{P_{(t)}\} - [k]\{u_i\} = m\{\ddot{u}_i\}. \tag{7.2}$$

For *free vibrations* $\{P_{(t)}\} = 0$, so that (7.2) becomes

$$[k]\{u_i\} + m\{\ddot{u}_i\} = 0. \tag{7.3}$$

If *simple harmonic motion* is assumed, then

$$\{u_i\} = \{A\}e^{j\omega t}$$

and

$$\{\ddot{u}_i\} = -\omega^2\{A\}e^{j\omega t} = -\omega^2\{u_i\} \tag{7.4}$$

where

$$\omega = \text{radian frequency, } j = \sqrt{-1}.$$

Substituting (7.4) into (7.3), the following is obtained:

$$([k] - \omega^2[m])\{u_i\} = 0, \tag{7.5}$$

and for the complete structure in global coordinates,

$$([K^\circ] - \omega^2[M^\circ])\{u_i^\circ\} = 0. \tag{7.6}$$

The condition $\{u_i^\circ\} = \{0\}$ is not of interest, therefore

$$([K^\circ] - \omega^2[M^\circ]) \text{ is singular,}$$

or

$$|[K^\circ] - \omega^2[M^\circ]| = 0. \tag{7.7}$$

For *constrained structures*, such as free-free beams, etc., $[K^\circ]$ is singular, and solution of the problem can be achieved by inverting $[M^\circ]$ and premultiplying it into (7.7) to give

$$|[M^\circ]^{-1}[K^\circ] - \omega^2\lceil I \rfloor| = 0. \tag{7.8}$$

Standard eigenvalue techniques can be used to solve (7.8), and normally, the smallest roots are of most importance.

For *constrained* or *over-constrained* structures, of which common types are 'statically determinate' and 'statically indeterminate' structures, respectively, either $[M^\circ]$ or $[K^\circ]$ can be inverted. If, however, the determination of the eigenvalues is by the 'Power' method [34], then it is usually best to invert $[K^\circ]$, and premultiply it into the equation, as follows:

$$(\lambda\lceil I \rfloor - [K^\circ]^{-1}[M^\circ])\{u_i^\circ\} = \{0\}. \tag{7.9}$$

In (7.9) it is usually necessary to determine the largest eigenvalues, where

$$\lambda = 1/\omega^2.$$

For very small problems, solution can be achieved by expanding the following determinant:

$$|[K^\circ] - \omega^2[M^\circ]| = 0. \tag{7.10}$$

For constrained and over-constrained structures, it is not necessary to use the entire $[K^o]$ and $[M^o]$ matrices in (7.9) and (7.10), as for these cases, 'constrained' versions can be used, as shown in (7.11) and (7.12):

$$\left| \lambda[\mathrm{I}] - [K_{11}]^{-1}[M_{11}] \right| = 0 \tag{7.11}$$

and

$$\left| [K_{11}] - \omega^2[M_{11}] \right| = 0 \tag{7.12}$$

where

$[K_{11}]$ is a smaller version of $[K^o]$, where the coefficients corresponding to the zero displacements, have been removed from $[K^o]$, as in section 2.4,

$[M_{11}]$ is a smaller version of $[M^o]$, where the coefficients corresponding to the zero displacements have been removed from $[M^o]$, in a similar manner to that for $[K_{11}]$.

7.2 DERIVATION OF MASS MATRICES

7.2.1 Virtual work method

Consider an infinitesimal element.

Dynamic force = mass × acceleration

$$= \rho\, \mathrm{d}(\mathrm{vol})\{\ddot{u}_{(x,y)}\}$$

$$= -\rho\omega^2\{u_{(x,y)}\}\,\mathrm{d}(\mathrm{vol})$$

Virtual work done by this force

$$= -\{\bar{u}_{(x,y)}\}^{\mathrm{T}}\rho\omega^2\{u_{(x,y)}\}\,\mathrm{d}(\mathrm{vol}),$$

and for the whole element,

Virtual work $= -\omega^2 \int\limits_{\mathrm{vol}} \{\bar{u}_{(x,y)}\}^{\mathrm{T}}\rho\{u_{(x,y)}\}\,\mathrm{d}(\mathrm{vol}).$

Now from (5.2):

$$\{u_{(x,y)}\} = [M_{(x,y)}]\{\alpha_i\},$$

and from (5.5):

$$\{u_i\} = [C]\{\alpha_i\},$$

therefore

$$\{\alpha\} = [C^{-1}]\{u_i\}.$$

Hence,

$$\{u_{(x,y)}\} = [M_{(x,y)}][C^{-1}]\{u_i\},$$

therefore

$$\text{Virtual work} = -\omega^2\{\bar{u}_i\}^T[C^{-1}]^T\int[M_{(x,y)}]^T$$
$$\times \rho[M_{(x,y)}]\,d(vol)[C^{-1}]\{u_i\}.$$

If this virtual work, done by the dynamic force, is added to (2.18), the following is obtained:

$$\{P_{(t)}\} = [k]\{u_i\} - \omega^2[m]\{u_i\},$$

which reveals that

$$[m] = [C^{-1}]^T[\tilde{m}][C^{-1}] \tag{7.13}$$

$$\{\tilde{m}\} = \int_{vol} [M_{(x,y)}]^T\rho[M_{(x,y)}]\,d(vol). \tag{7.14}$$

7.2.2 Method of minimum potential

An alternative method of finding $[m]$ is by the method of minimum potential.

$$\pi_p = U_e + \pi_k + W$$
$$= \tfrac{1}{2}\{u_i\}^T[k]\{u_i\} + \frac{1}{2}\int_{vol}\{\dot{u}_{(x,y)}\}^T\rho\{\dot{u}_{(x,y)}\}\,d(vol) - \{u_i\}^T\{p_i\}.$$

Now

$$\{u_{(x,y)}\} = [N]\{u_i\},$$

therefore

$$\{\dot{u}_{(x,y)}\} = [N]\{\dot{u}_i\} = [N]j\omega\{u_i\}.$$

Hence

$$\pi_p = \tfrac{1}{2}\{u_i\}^T[k]\{u_i\} - \tfrac{1}{2}\omega^2\int_{vol}\{u_i\}^T[N]^T\rho[N]\{u_i\}\,d(vol) - \{u_i\}^T\{P_{i(t)}\}$$
$$= \tfrac{1}{2}\{u_i\}^T[k]\{u_i\} - \tfrac{1}{2}\omega^2\{u_i^T\}[m]\{u_i\} - \{u_i\}^T\{P_{i(t)}\},$$

therefore

$$[m] = \int_{vol} [N]^T\rho[N]\,d(vol), \tag{7.15}$$

where

π_p = total potential energy
U_e = strain energy
π_k = kinetic energy
W = potential energy of load system.

7.3 PIN-JOINTED TRUSSES

7.3.1 Mass matrix for a plane rod (Fig. 7.1)

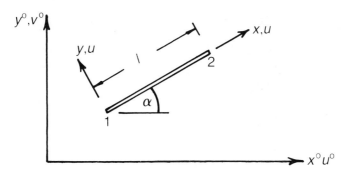

Fig. 7.1 – Plane rod element.

The assumed displacement function is the same as in equation (5.1). That is,

$$\{u_{(x, y)}\} = \alpha_1 + \alpha_2 x,$$

therefore

$$[M_{(x, y)}] = [1 \quad x].$$

Now,

$$[\tilde{m}] = \rho A \int_0^1 [M_{(x, y)}]^T [M_{(x, y)}] \, dx$$

$$= \rho A \int_0^1 \begin{bmatrix} 1 \\ x \end{bmatrix} [1 \quad x] \, dx$$

$$= \rho A \int_0^1 \begin{bmatrix} 1 & x \\ x & x^2 \end{bmatrix}$$

$$= \rho A \begin{bmatrix} l & l^2/2 \\ l^2/2 & l^3/3 \end{bmatrix}.$$

From (5.8),

$$[C]^{-1} = \begin{bmatrix} 1 & 0 \\ -1/l & 1/l \end{bmatrix},$$

therefore

$$[m] = \frac{\rho A l}{6} \begin{matrix} u_1 & u_2 \\ \begin{bmatrix} 2 & 1 \\ 1 & 2 \end{bmatrix} & \begin{matrix} u_1 \\ u_2 \end{matrix} \end{matrix}.$$

This is a mass matrix for a rod element in one dimension.

In *two dimensions*, the mass matrix for a plane rod is obtained by assuming the following displacement function:

$$\{u_{(x,y)}\} = \begin{Bmatrix} u \\ v \end{Bmatrix} = \begin{bmatrix} 1 & x & 0 & 0 \\ 0 & 0 & 1 & x \end{bmatrix} \begin{Bmatrix} \alpha_1 \\ \alpha_2 \\ \alpha_3 \\ \alpha_4 \end{Bmatrix} = [M_{(x,y)}]\{\alpha_i\} \tag{7.16}$$

and

$$[C] = \begin{matrix} & \begin{matrix} u_1 & v_1 & u_2 & v_2 \end{matrix} & \\ & \begin{bmatrix} 1 & 0 & 0 & 0 \\ 0 & 0 & 1 & 0 \\ 1 & l & 0 & 0 \\ 0 & 0 & 1 & l \end{bmatrix} & \begin{matrix} u_1 \\ v_1 \\ u_2 \\ v_2 \end{matrix} \end{matrix} \tag{7.17}$$

Substituting $[M(x, y)]$ from (7.16) and (7.17) into (7.13), it can be shown that the mass matrix for a *two-dimensional rod* is given by

$$[m] = \frac{\rho Al}{6} \begin{matrix} & \begin{matrix} u_1 & v_1 & u_2 & v_2 \end{matrix} & \\ & \begin{bmatrix} 2 & & \text{Symmetrical} & \\ 0 & 2 & & \\ 1 & 0 & 2 & \\ 0 & 1 & 0 & 2 \end{bmatrix} & \begin{matrix} u_1 \\ v_1 \\ u_2 \\ v_2 \end{matrix} \end{matrix} \tag{7.18}$$

7.3.2 Mass matrix for a space rod

Similarly, for the three-dimensional rod of Fig. 3.12, the assumed displacement function is given by

$$\{u_{(x,y)}\} = \begin{Bmatrix} u \\ v \\ w \end{Bmatrix} = \begin{bmatrix} 1 & x & 0 & 0 & 0 & 0 \\ 0 & 0 & 1 & x & 0 & 0 \\ 0 & 0 & 0 & 0 & 1 & x \end{bmatrix} \begin{Bmatrix} \alpha_1 \\ \alpha_2 \\ \alpha_3 \\ \alpha_4 \\ \alpha_5 \\ \alpha_6 \end{Bmatrix} = [M_{(x,y)}]\{\alpha_i\} \tag{7.19}$$

and

$$[C] = \begin{matrix} & \begin{matrix} u_1 & v_1 & w_1 & u_2 & v_2 & w_2 \end{matrix} & \\ & \begin{bmatrix} 1 & 0 & 0 & 0 & 0 & 0 \\ 0 & 0 & 1 & 0 & 0 & 0 \\ 0 & 0 & 0 & 0 & 1 & 0 \\ 1 & l & 0 & 0 & 0 & 0 \\ 0 & 0 & 1 & l & 0 & 0 \\ 0 & 0 & 0 & 0 & 1 & l \end{bmatrix} & \begin{matrix} u_1 \\ v_1 \\ w_1 \\ u_2 \\ v_2 \\ w_2 \end{matrix} \end{matrix} \tag{7.20}$$

Substituting $[M_{(x,\,y)}]$ from (7.19) and (7.20) into (7.13), it can be shown that the mass matrix for a *three-dimensional rod* is given by

$$[m] = \frac{\rho A l}{6} \begin{bmatrix} 2 & & & & & \\ 0 & 2 & \text{Symmetrical} & & & \\ 0 & 0 & 2 & & & \\ 1 & 0 & 0 & 2 & & \\ 0 & 1 & 0 & 0 & 2 & \\ 0 & 0 & 1 & 0 & 0 & 2 \end{bmatrix}. \tag{7.21}$$

7.3.3 Mass matrices in global coordinates

The mass matrix in global coordinates is given by

$$[m^\circ] = [\Xi]^T [m] [\Xi] \tag{7.22}$$

where

$[\Xi]$ is a matrix of directional cosines.

By substituting $[\Xi]$ from (3.8) and (7.18) into (7.22), it can be shown that the mass matrix for a *two-dimensional rod in global coordinates* is given by

$$
\begin{array}{cccc}
u_1^{\,\circ} & v_1^{\,\circ} & u_2^{\,\circ} & v_2^{\,\circ}
\end{array}
$$

$$[m^\circ] = \frac{\rho A l}{6} \begin{bmatrix} 2 & & \text{Symmetrical} & \\ 0 & 2 & & \\ 1 & 0 & 2 & \\ 0 & 1 & 0 & 2 \end{bmatrix} \begin{array}{c} u_1^{\,\circ} \\ v_1^{\,\circ} \\ u_2^{\,\circ} \\ v_2^{\,\circ} \end{array} \tag{7.23}$$

Similarly, by substituting (3.31) and (7.21) into (7.22), it can be shown that the mass matrix for a *three-dimensional rod in global coordinates* is given by

$$
\begin{array}{cccccc}
u_1^{\,\circ} & v_1^{\,\circ} & w_1^{\,\circ} & u_2^{\,\circ} & v_2^{\,\circ} & w_2^{\,\circ}
\end{array}
$$

$$[m^\circ] = \frac{\rho A l}{6} \begin{bmatrix} 2 & & & & & \\ 0 & 2 & & \text{Symmetrical} & & \\ 0 & 0 & 2 & & & \\ 1 & 0 & 0 & 2 & & \\ 0 & 1 & 0 & 0 & 2 & \\ 0 & 0 & 1 & 0 & 0 & 2 \end{bmatrix} \begin{array}{c} u_1^{\,\circ} \\ v_1^{\,\circ} \\ w_1^{\,\circ} \\ u_2^{\,\circ} \\ v_2^{\,\circ} \\ w_2^{\,\circ} \end{array} \tag{7.24}$$

7.3.4 Added masses

If in addition to the distributed masses, there is a concentrated mass of m_c at node 'i', then this additional mass is added to $[M^\circ]$ at node 'i', as follows:

$$u_i^\circ \quad v_i^\circ$$

$$[M^\circ] = [M^\circ] + m_c \begin{bmatrix} 1 & 0 \\ 0 & 1 \end{bmatrix} \begin{matrix} u_i^\circ \\ v_i^\circ \end{matrix}$$

for the *two-dimensional rod*, and

$$u_i^\circ \quad v_i^\circ \quad w_i^\circ$$

$$[M^\circ] = [M^\circ] + m_c \begin{bmatrix} 1 & 0 & 0 \\ 0 & 1 & 0 \\ 0 & 0 & 1 \end{bmatrix} \begin{matrix} u_i^\circ \\ v_i^\circ \\ w_i^\circ \end{matrix}$$

for the *three-dimensional rod*.

Similarly, if there are a number of additional concentrated masses at certain nodes, then these masses are added in the above manner at the appropriate nodes in the required number of 'directions', i.e. if the problem is a two-dimensional truss, then there are two directions at each node; and if the problem is a three-dimensional truss, there are three directions at each node.

7.3.5 Example 7.1

Determine the two natural frequencies of vibration for the plane pin-jointed truss of Fig. 3.6, given that $E = 2 \times 10^{11} \text{ N/m}^2$ and $\rho = 7860 \text{ kg/m}^3$. From 3.21,

$$u_i^\circ \qquad v_i^\circ$$

$$[K_{11}] = \frac{A_0 E}{10} \begin{bmatrix} 1.750 & 0.567 \\ 0.567 & 3.250 \end{bmatrix} \begin{matrix} u_i^\circ \\ v_i^\circ \end{matrix} . \tag{7.25}$$

To determine the mass matrices

Member 1–2

$l = 10 \text{ m} \qquad A = A_0.$

Substituting the above values into (7.23),

$$u_1^\circ \quad v_1^\circ \quad u_2^\circ \quad v_2^\circ$$

$$[m_{1-2}^\circ] = \frac{\rho A_0 \times 10}{6} \begin{bmatrix} 2 & & & \\ 0 & 2 & & \\ 1 & 0 & 2 & \\ 0 & 1 & 0 & 2 \end{bmatrix} \begin{matrix} u_1^\circ \\ v_1^\circ \\ u_2^\circ \\ v_2^\circ \end{matrix} . \tag{7.26}$$

Member 1–3

$l = 5 \text{ m} \qquad A = A_0.$

Substituting the above into (7.23),

$$[m_{1-3}^\circ] = \frac{\rho A_0 \times 5}{6} \begin{array}{cccc} u_1^\circ & v_1^\circ & u_3^\circ & v_3^\circ \\ \begin{bmatrix} 2 & & & \\ 0 & 2 & & \\ 1 & 0 & 2 & \\ 0 & 1 & 0 & 2 \end{bmatrix} & \begin{array}{l} u_1^\circ \\ v_1^\circ \\ u_3^\circ \\ v_3^\circ \end{array} \end{array} \qquad (7.27)$$

Member 1–4

$$l = 7.07 \text{ m} \qquad A = \sqrt{2}\,A_o$$

$$[m_{1-4}^\circ] = \frac{\rho A_0 \times 10}{6} \begin{array}{cccc} u_1^\circ & v_1^\circ & u_4^\circ & v_4^\circ \\ \begin{bmatrix} 2 & & & \\ 0 & 2 & & \\ 1 & 0 & 2 & \\ 0 & 1 & 0 & 2 \end{bmatrix} & \begin{array}{l} u_1^\circ \\ v_1^\circ \\ u_4^\circ \\ v_4^\circ \end{array} \end{array} \qquad (7.28)$$

Adding together those coefficients of (7.26) to (7.28); corresponding to the free displacements u_1° and v_1°, the following is obtained for the constrained mass matrix of the structure:

$$[M_{11}] = \frac{\rho A_0 \times 10}{6} \begin{array}{cc} u_1^\circ & v_1^\circ \\ \begin{bmatrix} 2 + 1 + 2 & 0 + 0 + 0 \\ 0 + 0 + 0 & 2 + 1 + 2 \end{bmatrix} & \begin{array}{l} u_1^\circ \\ v_1^\circ \end{array} \end{array}$$

$$[M_{11}] = \frac{\rho A_0 \times 10}{6} \begin{bmatrix} 5 & 0 \\ 0 & 5 \end{bmatrix}. \qquad (7.29)$$

Substituting (7.25) and (7.29) into (7.12), the following is obtained for the equation of motion:

$$\left| \frac{A_0 E}{10} \begin{bmatrix} 1.750 & 0.567 \\ 0.567 & 3.250 \end{bmatrix} - \omega^2 \frac{\rho A_o 10}{6} \begin{bmatrix} 5 & 0 \\ 0 & 5 \end{bmatrix} \right| = 0,$$

$$\left| \begin{array}{cc} (534351 - \omega^2) & 173130 \\ 173130 & (992366 - \omega^2) \end{array} \right| = 0. \qquad (7.30)$$

Expanding (7.30), the following is obtained:

$$\omega^4 - 1526717\,\omega^2 + 5 \times 10^{11} = 0,$$

from which

$$\omega^2 = \frac{1526717 \pm \sqrt{(1526712^2 - 4 \times 5 \times 10^{11})}}{2}$$

or

$$\omega_1 = 689.7\ \text{rad/s} \quad \text{and} \quad n_1 = 109.77\ \text{Hz} = \text{first natural frequency}$$

and

$$\omega_2 = 1025.2\ \text{rad/s} \quad \text{and} \quad n_2 = 163.2\ \text{Hz} = \text{second natural frequency.}$$

Check
Reference [44] tells us that the trace of a matrix is equal to the sum of its eigenvalues, so let us check this.

$$\omega_1{}^2 + \omega_2{}^2 = 1526721 = 534351 + 992366 = 1526717\ \text{OK.}$$

To obtain the *first eigenmode* (corresponding to ω_1), substitute the value of ω_1 into the 2nd row of (7.30), to give

$$173130\,u_1{}^\circ + 516680\,v_1{}^\circ = 0.$$

Let $u_1{}^\circ = 1$, hence, $v_1{}^\circ = -0.335$,

therefore

$$\text{1st eigenmode} = \lfloor u_1{}^\circ \quad v_1{}^\circ \rfloor = \lfloor 1 \quad -0.335 \rfloor.$$

To obtain the *2nd eigenmode* (corresponding to ω_2), substitute the value of ω_2 into the 1st row of (7.30), to give

$$-516684\,u_1{}^\circ + 173130\,v_1{}^\circ = 0.$$

Let $v_1{}^\circ = 1$, hence, $u_1{}^\circ = 0.335$,

therefore

$$\text{2nd eigenmode} = \lfloor u_1{}^\circ \quad v_1{}^\circ \rfloor = \lfloor 0.335 \quad 1 \rfloor.$$

N.B. The eigenmodes could have been obtained by substituting ω_1 into the first row, and ω_2 into the second row of (7.30), or by substituting both into any row.

7.3.6 Example 7.2

Determine the two lowest natural frequencies of vibration of Example 7.1, assuming that an additional mass of 100 kg is added to node 1 and that $A_0 = 0.001\ \text{m}^2$. From (7.25),

$$[K_{11}] = 2 \times 10^7 \begin{bmatrix} \overset{u_1{}^\circ}{1.750} & \overset{v_1{}^\circ}{0.567} \\ 0.567 & 3.250 \end{bmatrix} \begin{matrix} u_1{}^\circ \\ v_1{}^\circ \end{matrix} \qquad (7.31)$$

From (7.29),

$$[M_{11}] = 65.5 \begin{bmatrix} 1 & 0 \\ 0 & 1 \end{bmatrix} + 100 \begin{bmatrix} 1 & 0 \\ 0 & 1 \end{bmatrix}$$

$$= 165.5 \begin{matrix} u_1{}^\circ & v_1{}^\circ \\ \begin{bmatrix} 1 & 0 \\ 0 & 1 \end{bmatrix} & \begin{matrix} u_1{}^\circ \\ v_1{}^\circ \end{matrix} \end{matrix} \tag{7.32}$$

Substituting (7.31) and (7.32) into (7.12), the equation of motion is given by

$$\begin{vmatrix} 211480 - \omega^2 & 68520 \\ 68520 & 392750 - \omega^2 \end{vmatrix} = 0 \tag{7.33}$$

which on expansion, becomes

$$\omega^4 - 604230\,\omega^2 + 7.836 \times 10^{10} = 0,$$

the roots of which are

$$\underline{\omega_1 = 434.0 \text{ rad/s}} \quad \text{and} \quad \underline{n_1 = 69.1 \text{ Hz}}; \quad \text{and}$$

$$\underline{\omega_2 = 644.9 \text{ rad/s}} \quad \text{and} \quad \underline{n_2 = 102.6 \text{ Hz}.}$$

To obtain the *1st eigenmode* (corresponding to ω_1), substitute ω_1 into the 1st row of (7.33), as follows:

$$(211480 - 434^2)\,u_1{}^\circ + 68520\,v_1{}^\circ = 0.$$

Letting $u_1{}^\circ = 1$, then $v_1{}^\circ = -0.337$,

therefore

$$\text{1st eigenmode} = [u_1{}^\circ \quad v_1{}^\circ] = [1 \quad -0.337].$$

To obtain the *2nd eigenmode* (corresponding to ω_2), substitute ω_2 into the 2nd row of (7.33), as follows:

$$68520\,u_1{}^\circ + (392750 - 644.9^2)\,v_1{}^\circ = 0,$$

therefore

$$\text{2nd eigenmode} = [u_1{}^\circ \quad v_1{}^\circ] = [0.338 \quad 1].$$

N.B. The 1st eigenmode could have been obtained by substituting ω_1 into the 2nd row of (7.33), and the 2nd eigenmode by substituting ω_2 into the first row of (7.33).

7.3.7 Example 7.3

Determine the three lowest natural frequencies of vibration for the pin-jointed space truss of Fig. 3.13, given that $E = 2 \times 10^{11}$ N/m^2 and $\rho = 7860$ kg/m^3. From (3.37),

$$[K_{11}] = \frac{AE}{20} \begin{array}{ccc} u_4^{\,\circ} & v_4^{\,\circ} & w_4^{\,\circ} \\ \begin{bmatrix} 0.5 & 0 & 0 \\ 0 & 1 & 0.207 \\ 0 & 0.207 & 1.5 \end{bmatrix} & \begin{array}{c} u_4^{\,\circ} \\ v_4^{\,\circ} \\ w_4^{\,\circ} \end{array} \end{array} \qquad (7.34)$$

As all three members have the same value for ρ, A, and l, the constrained system mass matrix is given by

$$[M_{11}] = \rho Al \begin{array}{ccc} u_4^{\,\circ} & v_4^{\,\circ} & w_4^{\,\circ} \\ \begin{bmatrix} 1 & 0 & 0 \\ 0 & 1 & 0 \\ 0 & 0 & 1 \end{bmatrix} & \begin{array}{c} u_4^{\,\circ} \\ v_4^{\,\circ} \\ w_4^{\,\circ} \end{array} \end{array} \qquad (7.35)$$

Substituting (7.34) and (7.35) into (7.12), the equation of motion is given by

$$\left| \frac{AE}{20} \begin{bmatrix} 0.5 & 0 & 0 \\ 0 & 1 & 0.207 \\ 0 & 0.207 & 1.5 \end{bmatrix} - \omega^2 \rho A \times 20 \begin{bmatrix} 1 & 0 & 0 \\ 0 & 1 & 0 \\ 0 & 0 & 1 \end{bmatrix} \right| = 0$$

or

$$\begin{vmatrix} (31807 - \omega^2) & 0 & 0 \\ 0 & (63613 - \omega^2) & 13168 \\ 0 & 13168 & (95420 - \omega^2) \end{vmatrix} = 0. \qquad (7.36)$$

Expanding (7.36), the following cubic equation is obtained:

$$(31807 - \omega^2)[(63613 - \omega^2)(95420 - \omega^2) - 13168^2] = 0. \qquad (7.37)$$

From (7.37), it can readily be seen that

$$\underline{\omega_1 = \sqrt{31807} = 178.3 \text{ rad/s and } n_1 = 28.38 \text{ Hz}} = \text{1st frequency,}$$

and by inspection, the *1st eigenmode* corresponding to this eigenvalue

$$[u_4^{\,\circ} \, v_4^{\,\circ} \, w_4^{\,\circ}] = [1 \quad 0 \quad 0].$$

From (7.37), it can also be seen that

$$(63613 - \omega^2)(95420 - \omega^2) - 13168^2 = 0,$$

from which

$$\underline{\omega_2 = 242.7 \text{ rad/s}} \quad \text{and} \quad \underline{n_2 = 38.6 \text{ Hz}} = \text{2nd frequency,}$$

and

$$\underline{\omega_3 = 316.5 \text{ rad/s}} \quad \text{and} \quad \underline{n_3 = 50.37 \text{ Hz}} = \text{3rd frequency.}$$

To obtain the *2nd eigenmode* (corresponding to ω_2), substitute ω_2 into the second row of (7.36) and note that $u_4^\circ = 0$ for this case.

$$(63613 - 242.7^2)v_4^\circ + 13168w_4^\circ = 0,$$

therefore

$$2nd\ eigenmode = \lfloor u_4^\circ\ v_4^\circ\ w_4^\circ \rfloor = \lfloor 0 \quad 1 \quad -0.36 \rfloor.$$

To obtain the *3rd eigenmode*, substitute ω_3 into the third row of (7.36), and note that $u_4^\circ = 0$ for this case.

$$13168\,v_4^\circ + (95420 - 316.5^2)w_4^\circ = 0,$$

therefore

$$3rd\ eigenmode = \lfloor u_4^\circ\ v_4^\circ\ w_4^\circ \rfloor = \lfloor 0 \quad 0.36 \quad 1 \rfloor.$$

N.B. The 2nd eigenmode could have been determined by substituting ω_2 into the third row and vice versa for the third eigenmode.

7.3.8 Example 7.4

Determine the three lowest natural frequencies of vibration of Example 7.3, assuming that an additional concentrated mass of 500 kg is added to node 4 and that $A = 0.002$ m^2. From (7.34),

$$[K_{11}] = 2 \times 10^7 \begin{array}{ccc} u_4^\circ & v_4^\circ & w_4^\circ \\ \begin{bmatrix} 0.5 & 0 & 0 \\ 0 & 1 & 0.207 \\ 0 & 0.207 & 1.5 \end{bmatrix} & \begin{array}{c} u_4^\circ \\ v_4^\circ \\ w_4^\circ \end{array} \end{array}$$

and from (7.35),

$$[M_{11}] = 314.4 \begin{bmatrix} 1 & 0 & 0 \\ 0 & 1 & 0 \\ 0 & 0 & 1 \end{bmatrix} + 500 \begin{bmatrix} 1 & 0 & 0 \\ 0 & 1 & 0 \\ 0 & 0 & 1 \end{bmatrix}$$

$$= 814.4 \begin{bmatrix} 1 & 0 & 0 \\ 0 & 1 & 0 \\ 0 & 0 & 1 \end{bmatrix}$$

The dynamical equation becomes

$$\begin{vmatrix} (12279 - \omega^2) & 0 & 0 \\ 0 & (24558 - \omega^2) & 5083 \\ 0 & 5083 & (36837 - \omega^2) \end{vmatrix} = 0. \qquad (7.38)$$

From (7.38), the first eigenvalue is

$$\omega_1 = \sqrt{12279} = 110.8 \text{ rad/s} \quad \text{and} \quad n_1 = 17.64 \text{ Hz},$$

and by inspection the eigenmode corresponding to this eigenvalue is

$$\lfloor u_4{}^\circ \; v_4{}^\circ \; w_4{}^\circ \rfloor = \lfloor 1 \quad 0 \quad 0 \rfloor.$$

Similarly, from (7.38) the second and third eigenvalues are obtained from the following quadratic equation in ω^2.

$$\omega^4 - 61395 \, \omega^2 + 8.788 \times 10^8 = 0,$$

or

$$\omega_2 = 150.75 \text{ rad/s} \quad \text{and} \quad n_2 = 24 \text{ Hz}$$

and

$$\omega_3 = 196.64 \text{ rad/s} \quad \text{and} \quad n_2 = 31.3 \text{ Hz} .$$

The second and third eigenmodes, which correspond to ω_2 and ω_3 respectively, are as follows:

$$2nd \; eigenmode = \lfloor u_4{}^\circ \quad v_4{}^\circ \quad w_4{}^\circ \rfloor = \lfloor 0 \quad 1 \quad -0.36 \rfloor$$
$$3rd \; eigenmode = \lfloor u_4{}^\circ \quad v_4{}^\circ \quad w_4{}^\circ \rfloor = \lfloor 0 \quad 0.36 \quad 1 \rfloor.$$

7.4 BEAMS AND RIGID-JOINTED FRAMES

7.4.1 Mass matrix for a beam in flexure (Fig. 7.2)

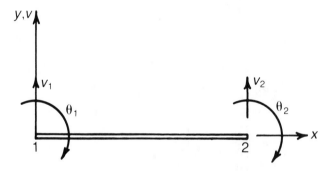

Fig. 7.2 – Plane beam element.

The assumed displacement function is

$$v = \alpha_1 + \alpha_2 x + \alpha_3 x^2 + \alpha_4 x^3,$$

which can be seen to give a linear distribution for bending moment, similar to the familiar beam element used in the displacement method (section 4.1).

Now, $\theta = -\dfrac{dv}{dx}$.

Hence, from considerations of the v and θ displacements,

$$[C] = \begin{bmatrix} 1 & 0 & 0 & 0 \\ 0 & -1 & 0 & 0 \\ 1 & l & l^2 & l^3 \\ 0 & -1 & -2l & -3l^2 \end{bmatrix}$$

and

$$[C]^{-1} = \begin{bmatrix} 1 & 0 & 0 & 0 \\ 0 & -1 & 0 & 0 \\ -3/l^2 & 2/l & 3/l^2 & 1/l \\ 2/l^3 & -1/l^2 & -2/l^3 & -1/l^2 \end{bmatrix} . \tag{7.39}$$

Now, $[M_{(x,\,y)}] = \begin{bmatrix} 1 & x & x^2 & x^3 \end{bmatrix}$

and

$$[\tilde{m}] = \rho \int_{\text{vol}} [M_{(x,\,y)}]^{\mathsf{T}} [M_{(x,\,y)}] \, d(\text{vol}),$$

therefore

$$[\tilde{m}] = \rho A \int_{0}^{1} \begin{bmatrix} 1 \\ x \\ x^2 \\ x^3 \end{bmatrix} \begin{bmatrix} 1 & x & x^2 & x^3 \end{bmatrix} dx$$

$$= \rho A \begin{bmatrix} l & l^2/2 & l^3/3 & l^4/4 \\ l^2/2 & l^3/3 & l^4/4 & l^5/5 \\ l^3/3 & l^4/4 & l^5/5 & l^6/6 \\ l^4/4 & l^5/5 & l^6/6 & l^7/7 \end{bmatrix} . \tag{7.40}$$

Now,

$$[m] = [C^{-1}]^{\mathsf{T}} [\tilde{m}] [C^{-1}],$$

therefore from (7.39) and (7.40),

$$[m] = \frac{\rho A l}{420} \begin{array}{cccc} v_1 & \theta_1 & v_2 & \theta_2 \\ \begin{bmatrix} 156 & & & \\ -22l & 4l^2 & & \text{sym} \\ 54 & -13l & 156 & \\ 13l & -3l^2 & 22l & 4l^2 \end{bmatrix} & \begin{array}{c} v_1 \\ \theta_1 \\ v_2 \\ \theta_2 \end{array} \end{array} . \tag{7.41}$$

7.4.2 Example 7.5

Calculate the smallest natural frequency of vibration for a uniform section cantilever, using one element model.

The element and system stiffness matrices are given by

$$[k_{1-2}] = [K] = EI \begin{array}{cc} \begin{array}{cccc} v_1 & \theta_1 & v_2 & \theta_2 \end{array} \\ \begin{bmatrix} 12/l^3 & & & \\ -6/l^2 & 4/l & & \\ -12/l^3 & 6/l^2 & 12/l^3 & \\ -6/l^2 & 2/l & 6/l^2 & 4/l \end{bmatrix} \begin{array}{c} v_1 \\ \theta_1 \\ v_2 \\ \theta_2 \end{array} \end{array} \qquad (7.42)$$

and the corresponding mass matrices are given by

$$[m_{1-2}] = [M] = \frac{\rho A l}{6} \begin{array}{cc} \begin{array}{cccc} v_1 & \theta_1 & v_2 & \theta_2 \end{array} \\ \begin{bmatrix} 156 & & & \\ -22l & 4l^2 & & \\ 54 & -13l & 156 & \\ 13l & -3l^2 & 22l & 4l^2 \end{bmatrix} \begin{array}{c} v_1 \\ \theta_1 \\ v_2 \\ \theta_2 \end{array} \end{array} \qquad (7.43)$$

As the cantilever is fixed at one node (say node 1), the displacements corresponding to this node must be suppressed, so that only those coefficients from (7.42) and (7.43) which correspond to the free displacements need be considered for $[K_{11}]$ and $[M_{11}]$, i.e.,

$$[K_{11}] = EI \begin{array}{cc} \begin{array}{cc} v_2 & \theta_2 \end{array} \\ \begin{bmatrix} 12/l^3 & 6/l^2 \\ 6/l^2 & 4/l \end{bmatrix} \begin{array}{c} v_2 \\ \theta \end{array} \end{array} \qquad (7.44)$$

and

$$[M_{11}] = \frac{\rho A l}{420} \begin{array}{cc} \begin{array}{cc} v_2 & \theta_2 \end{array} \\ \begin{bmatrix} 156 & 22l \\ 22l & 4l^2 \end{bmatrix} \begin{array}{c} v_2 \\ \theta_2 \end{array} \end{array} \qquad (7.45)$$

Substituting (7.44) and (7.45) into the dynamical equation, the following is obtained:

$$\left| EI \begin{bmatrix} 12l^3 & 6/l^2 \\ 6/l^2 & 4/l \end{bmatrix} - \omega^2 \frac{\rho A l}{420} \begin{bmatrix} 156 & 22l \\ 22l & 4l^2 \end{bmatrix} \right| = 0.$$

Now,

$$[K_{11}]^{-1} = \frac{1}{EI} \begin{bmatrix} l^3/3 & -l^2/2 \\ -l^2/2 & l \end{bmatrix},$$

therefore

$$\left(\lambda \begin{bmatrix} 1 & 0 \\ 0 & 1 \end{bmatrix} - \frac{\rho Al}{420 EI} \begin{bmatrix} l^3/3 & -l^2/2 \\ -l^2/2 & l \end{bmatrix} \begin{bmatrix} 156 & 22l \\ 221 & 4l^2 \end{bmatrix} \right) \begin{Bmatrix} v_2 \\ \theta_2 \end{Bmatrix} = \begin{Bmatrix} 0 \\ 0 \end{Bmatrix}$$

or

$$\begin{vmatrix} \dfrac{41\rho Al^4}{420 EI} - \lambda & \dfrac{16\rho Al^5}{1260 EI} \\ -\dfrac{56\rho Al^3}{420 EI} & -\dfrac{\rho Al^4}{60 EI} - \lambda \end{vmatrix} = 0$$

$$\lambda = 0.0801 \, \rho Al^4/(EI),$$

but

$$\lambda = \frac{1}{\omega^2}$$

therefore

$$\omega = \frac{3.53}{l^2} \sqrt{\frac{EI}{\rho A}} \tag{7.46}$$

with an eigenmode $= \lfloor v_2 \quad \theta_2 \rfloor = \lfloor -0.73 \quad 1 \rfloor$.

It is interesting to note that this comparatively simple idealisation overestimates the exact value by only about 0.5 per cent.

To determine the elemental mass matrix for a *beam in global coordinates*, a similar process is adopted to that for the stiffness matrix. That is,

$$[m_b{}^o] = [\Xi]^T [m] [\Xi]$$

where $[\Xi]$ is obtained from section 4.3.

Using the above expressions, it can be shown that for a beam in global coordinates, the elemental mass matrix is given by

$$[m_b^o] = \frac{\rho Al}{420} \begin{bmatrix} 156s^2 & & & & & \\ -156cs & 156c^2 & & & & \\ 22ls & -22lc & 4l^2 & & & \\ 54s^2 & -54cs & 13ls & 156s^2 & & \\ -54cs & 54c^2 & -13lc & -156cs & 156c^2 & \\ -13ls & 13lc & -3l^2 & -22ls & 22lc & 4l^2 \end{bmatrix} \begin{matrix} u_1{}^o \\ v_1{}^o \\ \theta_1 \\ u_2{}^o \\ v_2{}^o \\ \theta_2. \end{matrix}$$

$$\begin{matrix} u_1{}^o & v_1{}^o & \theta_1 & u_2{}^o & v_2{}^o & \theta_2 \end{matrix}$$

$$\tag{7.47}$$

7.4.3 Rigid-jointed plane frame element

This element can be obtained by superimposing (7.47) with the appropriate parts of a rod element, as in (7.48).

$$[m^o] = [m_r{}^o] + [m_b{}^o] \tag{7.48}$$

where

$$[m_r^\circ] = [\Xi]^T[m_r][\Xi]$$

$$= [\Xi]^T \frac{\rho Al}{6} \begin{array}{c} \begin{array}{cccccc} u_1 & v_1 & \theta_1 & u_2 & v_2 & \theta_2 \end{array} \\ \begin{bmatrix} 2 & 0 & 0 & 1 & 0 & 0 \\ 0 & 0 & 0 & 0 & 0 & 0 \\ 0 & 0 & 0 & 0 & 0 & 0 \\ 1 & 0 & 0 & 2 & 0 & 0 \\ 0 & 0 & 0 & 0 & 0 & 0 \\ 0 & 0 & 0 & 0 & 0 & 0 \end{bmatrix} \begin{array}{c} u_1 \\ v_1 \\ \theta_1 \\ u_2 \\ v_2 \\ \theta_2 \end{array} \end{array} [\Xi]$$

$$[m_r^\circ] = \frac{\rho Al}{6} \begin{array}{c} \begin{array}{cccccc} u_1^\circ & v_1^\circ & \theta_1 & u_2^\circ & v_2^\circ & \theta_2 \end{array} \\ \begin{bmatrix} 2c^2 & & & & & \\ 2cs & 2s^2 & & \text{Symmetrical} & & \\ 0 & 0 & 0 & & & \\ c^2 & cs & 0 & 2c^2 & & \\ cs & s^2 & 0 & 2cs & 2s^2 & \\ 0 & 0 & 0 & 0 & 0 & 0 \end{bmatrix} \begin{array}{c} u_1^\circ \\ v_1^\circ \\ \theta_1 \\ u_2^\circ \\ v_2^\circ \\ \theta_2 \end{array} \end{array} \qquad (7.49)$$

It should be noted that $[m_r]$ contains no coefficients of mass in the 'v' direction, because this effect has already been included in $[m_b]$.

7.4.4 Torque bar

The mass matrix for a prismatic torque bar is now derived. From Fig. 7.3, the angle of twist 'ϕ' is given by

$$\phi = [(1-\xi) \quad \xi]\begin{Bmatrix} \phi_1 \\ \phi_2 \end{Bmatrix} = [N]\{u_i\}$$

Fig. 7.3.

where $\xi = x/l$, but as the rotational displacement, at any point ξ, varies with 'r'

$$[m] = \int_{\text{vol}} [N]^T r\rho[N] r \, d(\text{vol})$$

$$= \frac{\rho I_x l}{6} \begin{array}{cc} \phi_1 & \phi_2 \\ \begin{bmatrix} 2 & 1 \\ 1 & 2 \end{bmatrix} & \begin{array}{c} \phi_1 \\ \phi_2, \end{array} \end{array} \tag{50}$$

where, I_x = 2nd polar moment of area of the cross-section of the torque bar about its centroid and in the direction of the 'x' axis.

7.4.5 Additional concentrated masses

If a beam or a rigid-jointed plane frame has a number of additional masses at certain nodes, then the mass matrix of the structure must have added to it the magnitudes of these additional masses, together with their corresponding mass moments of inertia, to the appropriate nodes. For example, for a *beam*, with an additional mass at node 'i', the following modification must be carried out:

$$[M] = [M] + \begin{array}{cc} v_i & \theta_i \\ \begin{bmatrix} m_c & 0 \\ 0 & I_c \end{bmatrix} & \begin{array}{c} v_i \\ \theta_i. \end{array} \end{array} \tag{7.51}$$

Similarly, *for a rigid-jointed plane frame*, with an additional mass at node 'i', equation (7.51) will become

$$[M] = [M] + \begin{array}{ccc} u_i^\circ & v_i^\circ & \theta_i \\ \begin{bmatrix} m_c & 0 & 0 \\ 0 & m_c & 0 \\ 0 & 0 & I_c \end{bmatrix} & \begin{array}{c} u_i^\circ \\ v_i^\circ \\ \theta_i, \end{array} \end{array} \tag{7.52}$$

where
$\quad m_c$ = additional concentrated mass at node 'i'

$\quad I_c$ = the mass moment of inertia of this mass at node 'i'.

7.4.6 Example 7.6

Determine the lowest natural frequency of the cantilever of Example 7.5, assuming that there is, added to the free node, an additional mass of 100 kg, with a corresponding mass moment of inertia of 2 kg m^2.

The following values may also be assumed

$$l = 2\,\text{m} \quad A = 0.00785\,\text{m}^2 \quad I = 4.909 \times 10^{-6}\,\text{m}^4$$
$$E = 2 \times 10^{11}\,\text{N/m}^2 \quad \rho = 7860\,\text{kg/m}^3.$$

Without the added mass, and from (7.46),

$$\omega = \frac{3.53}{4} \sqrt{\left(\frac{2 \times 10^{11} \times 4.909 \times 10^{-6}}{7860 \times 0.00785} \right)}$$

$$\underline{\omega = 111.32\,\text{rad/s}} \quad \text{and} \quad \underline{n = 17.72\,\text{Hz}}.$$

With *added mass*, the dynamic equation becomes

$$\left| 981800 \begin{bmatrix} 1.5 & 1.5 \\ 1.5 & 2 \end{bmatrix} - \omega^2 \begin{bmatrix} 45.84 & 12.93 \\ 12.93 & 4.70 \end{bmatrix} + \begin{bmatrix} 100 & 0 \\ 0 & 2 \end{bmatrix} \right| = 0$$

Now,

$$[K_{11}]^{-1} = \frac{1}{736350} \begin{bmatrix} 2 & -1.5 \\ -1.5 & 1.5 \end{bmatrix} \tag{7.53}$$

therefore

$$\left| \lambda[I] - \frac{1}{736350} \begin{bmatrix} 2 & -1.5 \\ -1.5 & 1.5 \end{bmatrix} \begin{bmatrix} 145.84 & 12.93 \\ 12.93 & 6.7 \end{bmatrix} \right| = 0, \tag{7.54}$$

from which the following quadratic is obtained:

$$\lambda^2 = -3.571 \times 10^{-4} \lambda + 1.118 \times 10^{-9} = 0$$
$$\lambda_1 = 3.539 \times 10^{-4}$$
$$\omega_1 = \sqrt{1/\lambda_1} = 53.15 \, \text{rad/s} \quad \text{and} \quad n_1 = 8.46 \, \text{Hz}.$$

To obtain the *first eigenmode*, substitute into the first equation of 7.54, to give

$$1.59 \times 10^{-5} v_2 + 2.147 \times 10^{-5} \theta_2 = 0,$$

i.e., 1st eigenmode $= \lfloor v_2 \quad \theta_2 \rfloor = \lfloor 1 \quad -0.74 \rfloor$.

7.4.7

To determine the natural frequencies of a rigid-jointed plane frame without the aid of a computer will be quite a tedious task, so the ship's deckhouse of Fig. 7.4 will be analysed by the computer program "VIBRJPF" of reference [15].

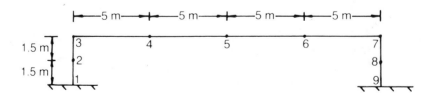

Fig. 7.4 – Ship's deckhouse.

The details of the framework are as follows:

$$E = 6.67 \times 10^{10} \, \text{N/m}^2$$
$$\rho = 2620 \, \text{kg/m}^3.$$

Vertical members

$$I = 4.19 \times 10^{-6}\, \text{m}^4$$
$$A = 2.4 \times 10^{-3}\, \text{m}^2.$$

Horizontal member

$$I = 6.491 \times 10^{-5}\, \text{m}^4$$
$$A = 6.8 \times 10^{-3}\, \text{m}^2.$$

The eigenmodes, together with the natural frequencies (n), are shown in Fig. 7.5, where it can be seen that the first eigenmode is a beam mode, and the second a sway mode. The third and fourth modes are frame modes, the former being antisymmetric and the latter being symmetric. It can be seen that all these modes are well within the exciting frequencies that normally exist on a ship.

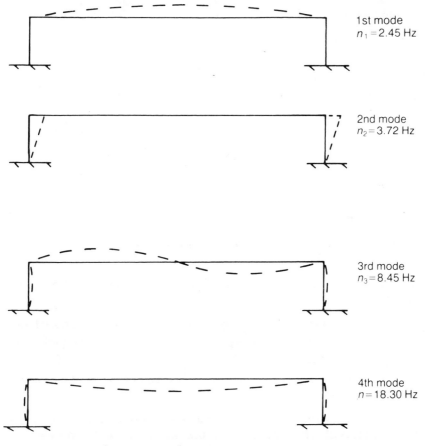

1st mode
$n_1 = 2.45$ Hz

2nd mode
$n_2 = 3.72$ Hz

3rd mode
$n_3 = 8.45$ Hz

4th mode
$n = 18.30$ Hz

Fig. 7.5 – Natural frequencies and eigenmodes of a ship's deckhouse.

7.4.8

Whereas the free vibration analysis of a rigid-jointed plane frame, without the aid of a digital computer, would prove to be a very tedious task, the dynamic analysis of most practical rigid-jointed space frames would be virtually impossible without the aid of a digital computer. This is because there are six degrees of freedom per mode, and for any practical structure, such analysis would require matrix reduction, as carried out in reference [46] and described in section 7.9.

7.5 AREA COORDINATES

This is a natural coordinate system used for triangles. It is particularly useful if integrals are required of various functions over a triangle.

Prior to describing the method, it will be necessary to define the three coordinates that are used in this system. These coordinates are L_1, L_2, and L_3, as shown in Fig. 7.6, where it can be seen that they start from the sides opposite

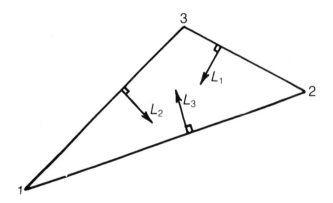

Fig. 7.6 – Area coordinates for a triangle.

their respective nodal numbers. Each of the coordinates varies from 0 to 1 as follows:

$L_1 = 0$ on side 2–3 and increases to a maximum value of 1 at node 1 (or i).

$L_2 = 0$ on side 1–3 and increases to a maximum value of 1 at node 2 (or j), as shown in Fig. 7.7.

$L_3 = 0$ on side 1–2 and increases to a maximum value of 1 at node 3 (or k).

L_1, L_2, and L_3 are called area coordinates, because they are defined by the ratios of areas of sub-triangles to the total area of the whole triangle.

Consider any point 'p' in the triangle of Fig. 7.8. This point can be defined

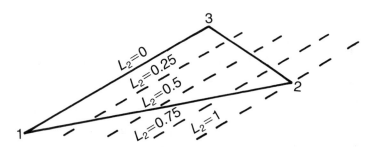

Fig. 7.7 – Variation of L_2.

by the area coordinates L_1, L_2, and L_3, which are as follows:

$$L_1 = A_1/\Delta$$
$$L_2 = A_2/\Delta \qquad\qquad (7.55)$$
$$L_3 = A_3/\Delta,$$

where

A_1, A_2, and A_3 = areas of sub-triangle (1), (2), and (3).

Δ = area of whole triangle.

To calculate L_1
Consider the triangle of Fig. 7.8.

$$\Delta = B_1 H_1/2$$
$$A_1 = \text{area of sub-triangle } ①$$

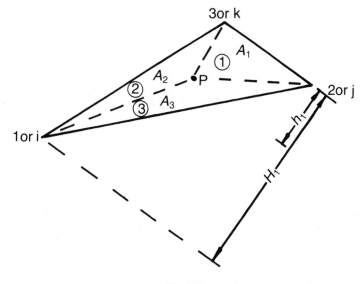

Fig. 7.8.

then

$$L_1 = \frac{A_1}{\Delta} = \frac{B_1 h_1}{2} \times \frac{2}{B_1 H_1},$$

therefore

$$\underline{L_1 = h_1/H_1}$$

Similarly, if $h_2, h_3, H_2,$ and H_3 represent similar dimensions with respect to sub-triangles ② and ③, then

$$L_2 = h_2/H_2$$

and

$$L_3 = h_3/H_3$$

Now, as $\Delta = A_1 + A_2 + A_3,$

$$1 = \frac{A_1 + A_2 + A_3}{\Delta},$$

or,

$$\underline{1 = L_1 + L_2 + L_3.} \tag{7.56}$$

It is evident that, as

at node 1, $L_1 = 1$ and $L_2 = L_3 = 0$
and at node 2, $L_2 = 1$ and $L_1 = L_3 = 0$
and at node 3, $L_3 = 1$ and $L_1 = L_2 = 0,$

the co-ordinates x and y of any point in the triangle can be represented by

$$x = L_1 x_i + L_2 x_j + L_3 x_k$$
$$y = L_1 y_i + L_2 y_j + L_3 y_k \tag{7.57}$$

Now the main reason for the popularity of area coordinates is because many of the integals associated with them are given by simple expressions, as follows:

$$\int_L L_1{}^a L_2{}^b \, \mathrm{d}L = \frac{a! \, b!}{(a+b+1)!} L \tag{7.58}$$

where $L = $ the length of the side over which the integral is carried out, and

$$\int_A L_1{}^a L_2{}^b L_3{}^c \, \mathrm{d}A = \frac{a! \, b! \, c!}{(a+b+c+2)!} 2\Delta \tag{7.59}$$

where $\mathrm{d}A = $ elemental area of triangle.

N.B. It must be remembered that $0! = 1.$

7.6 MASS MATRICES FOR IN-PLANE ELEMENTS

7.6.1

Now from (7.57)

$$
\begin{Bmatrix} x \\ y \end{Bmatrix} = \begin{bmatrix} L_1 & L_2 & L_3 & 0 & 0 & 0 \\ 0 & 0 & 0 & L_1 & L_2 & L_3 \end{bmatrix} \begin{Bmatrix} x_i^{\,o} \\ y_i^{\,o} \\ x_j^{\,o} \\ y_j^{\,o} \\ x_k^{\,o} \\ y_k^{\,o} \end{Bmatrix}
$$

and

$$
\begin{Bmatrix} u \\ v \end{Bmatrix} = \begin{bmatrix} L_1 & L_2 & L_3 & 0 & 0 & 0 \\ 0 & 0 & 0 & L_1 & L_2 & L_3 \end{bmatrix} \begin{Bmatrix} u_i^{\,o} \\ v_i^{\,o} \\ u_j^{\,o} \\ v_j^{\,o} \\ u_k^{\,o} \\ v_k^{\,o} \end{Bmatrix}
$$

$$
= [N]\{u_i^{\,o}\}.
$$

Now,

$$
\begin{aligned}
[m^o] &= \int [N]^T \rho [N] \, \mathrm{d}(\mathrm{vol}) \\
&= \rho t \int [N]^{\mathrm{T}} [N] \, \mathrm{d}A
\end{aligned}
$$

$$
= \rho t \int \begin{bmatrix} L_1 & 0 \\ L_2 & 0 \\ L_3 & 0 \\ 0 & L_1 \\ 0 & L_2 \\ 0 & L_3 \end{bmatrix} \begin{bmatrix} L_1 & L_2 & L_3 & 0 & 0 & 0 \\ 0 & 0 & 0 & L_1 & L_2 & L_3 \end{bmatrix} \mathrm{d}A
$$

$$
= \rho t \begin{bmatrix} L_1^{\,2} & L_1 L_2 & L_1 L_3 & 0 & 0 & 0 \\ & L_2^{\,2} & L_2 L_3 & 0 & 0 & 0 \\ & & L_3^{\,2} & 0 & 0 & 0 \\ & & & L_1^{\,2} & L_1 L_2 & L_1 L_3 \\ & & & & L_2^{\,2} & L_2 L_3 \\ & & & & & L_3^{\,2} \end{bmatrix}
$$

Now from (7.59),

$$[m^\circ] = \frac{\rho\Delta t}{12} \begin{array}{c} \begin{array}{cccccc} u_1{}^\circ & u_2{}^\circ & u_3{}^\circ & v_1{}^\circ & v_2{}^\circ & v_3{}^\circ \end{array} \\ \left[\begin{array}{ccc|ccc} 2 & 1 & 1 & 0 & 0 & 0 \\ 1 & 2 & 1 & 0 & 0 & 0 \\ 1 & 1 & 2 & 0 & 0 & 0 \\ \hline 0 & 0 & 0 & 2 & 1 & 1 \\ 0 & 0 & 0 & 1 & 2 & 1 \\ 0 & 0 & 0 & 1 & 1 & 2 \end{array}\right] \end{array} \begin{array}{c} u_1{}^\circ \\ u_2{}^\circ \\ u_3{}^\circ \\ v_1{}^\circ \\ v_2{}^\circ \\ v_3{}^\circ \end{array}, $$

which when rearranged becomes

$$[m^\circ] = \frac{\rho\Delta t}{12} \begin{array}{c} \begin{array}{cccccc} u_1 & v_1 & u_2 & v_2 & u_3 & v_3 \end{array} \\ \left[\begin{array}{cccccc} 2 & 0 & 1 & 0 & 1 & 0 \\ & 2 & 0 & 1 & 0 & 1 \\ & & 2 & 0 & 1 & 0 \\ \text{Symmetrical} & & & 2 & 0 & 1 \\ & & & & 2 & 0 \\ & & & & & 2 \end{array}\right] \end{array} \begin{array}{c} u_1 \\ v_1 \\ u_2 \\ v_2 \\ u_3 \\ v_3 \end{array}. \tag{7.60}$$

7.6.2 In-plane axisymmetric annular element

This problem is axisymmetric, i.e. the radial displacement is constant for any particular value of radius 'r'. Hence, it will be convenient to define the element with two nodal circles, one internal and one external, as shown in Fig. 7.9.

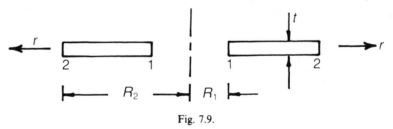

Fig. 7.9.

Now, from Section 5.2.4,

$$[N] = \frac{1}{(R_2 - R_1)} \left[(R_2 - r)\ (r - R_1) \right].$$

Now,

$$[m] = \int [N]^T \rho [N]\, 2\pi r\, dr\, t$$

$$= \frac{2\pi \rho t}{(R_2 - R_1)^2} \int_{R_1}^{R_2} \begin{bmatrix} (R_2 - r) \\ (r - R_1) \end{bmatrix} \left[(R_2 - r)\ (r - R_1) \right] r\, dr$$

$$= \frac{2\pi \rho t}{(R_2 - R_1)^2} \begin{bmatrix} m_{11} & m_{12} \\ m_{21} & m_{22} \end{bmatrix}$$

where

$$m_{11} = -\frac{1}{12}(3R_1{}^4 - 8R_1{}^3 R_2 + 6R_1{}^2 R_2{}^2 - R_2{}^4)$$

$$m_{12} = m_{21} = -\frac{1}{12}(R_1{}^4 - 2R_1{}^3 R_2 + 2R_1 R_2{}^3 - R_2{}^4)$$

$$m_{22} = -\frac{1}{12}(R_1{}^4 - 6R_1{}^2 R_2{}^2 + 8R_1 R_2{}^3 - 3R_2{}^4).$$

7.7 MASS MATRICES FOR TRIANGULAR AND QUADRILATERAL PLATES IN BENDING

Four out-of-plane plate bending elements will be described in this section, two non-conforming elements and two partially conforming ones, and comparisons will be made between three of them.

7.7.1 Non-triangular plate bending element (Clought & Tocher)

From Equation (5.71),

$$[M_{(x,\,y)}] = [1 \quad x \quad y \quad x^2 \quad xy \quad y^2 \quad x^3 \quad (x^2 y + y^2 x)y^3].$$

Now, $[\tilde{m}] = \rho t \int \int [M_{(x,\,y)}]^{\mathrm{T}} [M_{(x,\,y)}] \, dx \, dy$

therefore

$$[\tilde{m}] = \int_A \rho t \begin{bmatrix} 1 & & & \\ x & x^2 & & \\ y & xy & y^2 & \\ x^2 & x^3 & x^2 y & x^4 \\ xy & x^2 y & xy^2 & x^3 y \\ y^2 & xy^2 & y^3 & x^2 y^2 \\ x^3 & x^4 & x^3 y & x^5 \\ (x^2 y + xy^2) & (x^3 y + x^2 y^2) & (x^2 y^2 + xy^3) & (x^4 y + x^3 y^2) \\ y^3 & xy^3 & y^4 & x^2 y^3 \end{bmatrix}$$

$$\begin{bmatrix} x^2 y^2 & & & \\ xy^3 & y^4 & & \\ x^4 y & x^3 y^2 & x^6 & \\ (x^3 y^2 + x^2 y^3) & (x^2 y^3 + xy^4) & (x^5 y + x^4 y^2) & (x^2 y + xy^2)^2 \\ xy^4 & y^5 & x^3 y^3 & (x^2 y^4 + xy^5) \quad y^6 \end{bmatrix} \, dA \qquad (7.61)$$

The integrals of (7.61) were obtained with the assistance of a Jacobian transformation, where the x–y coordinates were transferred to the ξ–η coordinates of Figs 7.10 and 7.11, as suggested by Przemieniecki [45].

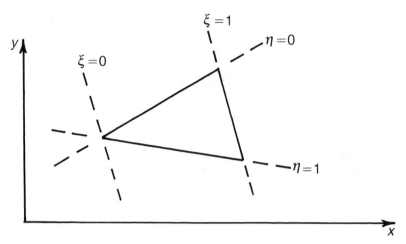

Fig. 7.10.

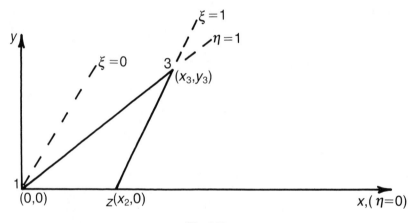

Fig. 7.11.

7.7.2

Some of these integrals are given in Chapter 5, and the remainder are as follows

$$\int_A y^3 \, dA = x_2 y_3^4 / 20$$

$$\int_A y^4 \, dA = x_2 y_3^5 / 30$$

$$\int_A y^5 \, dA = x_2 y_3^6 / 42$$

$$\int_A y^6 \, dA = x_2 y_3^7 / 56$$

$$\int_A x^3 \, dA = x_2 y_3 (x_3{}^3 + x_2 x_3{}^2 + x_2{}^2 x_3 + x_2{}^3)/20$$

$$\int_A x^4 \, dA = x_2 y_3 (x_3{}^4 + x_2 x_3{}^3 + x_2{}^2 x_3{}^2 + x_2{}^3 x_3 + x_2{}^4)/30$$

$$\int_A x^5 \, dA = x_2 y_3 (x_3{}^5 + x_2 x_3{}^4 + x_2{}^2 x_3{}^3 + x_2{}^3 x_3{}^2 + x_2{}^4 x_3 + x_2{}^5)/42$$

$$\int_A x^6 \, dA = x_2 y_3 (x_3{}^6 + x_2 x_3{}^5 + x_2{}^2 x_3{}^4 + x_2{}^3 x_3{}^3 + x_2{}^4 x_3{}^2$$
$$+ x_2{}^5 x_3 + x_2{}^6)/56$$

$$\int_A y^2 x \, dA = x_2 y_3{}^3 (x_2 + 3 x_3)/60$$

$$\int_A y^3 x \, dA = x_2 y_3{}^4 (x_2 + 4 x_3)/120$$

$$\int_A y^4 x \, dA = x_2 y_3{}^5 (x_2 + 5 x_3)/210$$

$$\int_A y^5 x \, dA = x_2 y_3{}^6 (x_2 + 6 x_3)/336$$

$$\int_A y x^2 \, dA = x_2 y_3{}^2 (3 x_3{}^2 + 2 x_2 x_3 + x_2{}^2)/60$$

$$\int_A y^2 x^2 \, dA = x_2 y_3{}^3 (6 x_3{}^2 + 3 x_2 x_3 + x_2{}^2)/180$$

$$\int_A y^3 x^2 \, dA = x_2 y_3{}^4 (10 x_3{}^2 + 4 x_2 x_3 + x_2{}^2)/420$$

$$\int_A y^4 x^2 \, dA = x_2 y_3{}^5 (15 x_3{}^2 + 5 x_2 x_3 + x_2{}^2)/840$$

$$\int_A y x^3 \, dA = x_2 y_3{}^2 (4 x_3{}^3 + 3 x_2 x_3{}^2 + 2 x_2{}^2 x_3 + x_2{}^3)/120$$

$$\int_A y^2 x^3 \, dA = x_2 y_3{}^3 (10 x_3{}^3 + 6 x_2 x_3{}^2 + 3 x_2{}^2 x_3 + x_2{}^3)/420$$

$$\int_A y^3 x^3 \, dA = x_2 y_3{}^4 (20 x_3{}^3 + 10 x_2 x_3{}^2 + 4 x_2{}^2 x_3 + x_2{}^3)/1120$$

$$\int_A y x^4 \, dA = x_2 y_3{}^2 (5 x_3{}^4 + 4 x_2 x_3{}^3 + 3 x_2{}^2 x_3{}^2 + 2 x_2{}^3 x_3 + x_2{}^4)/210$$

$$\int_A y^2 x^4 \, dA = x_2 y_3{}^3 (15 x_3{}^4 + 10 x_2 x_3{}^3 + 6 x_2{}^2 x_3{}^2 + 3 x_2{}^3 x_3 + x_2{}^4)/840$$

$$\int_A y x^5 \, dA = x_2 y_3{}^2 (6 x_3{}^5 + 5 x_2 x_3{}^4 + 4 x_2{}^2 x_3{}^3 + 3 x_2{}^2 x_3{}^2$$
$$+ 2 x_2{}^4 x_3 + x_2{}^5)/336$$

Alternatively, the integrals within the triangular boundary 1–2–3 of Fig. 7.11 could have been obtained directly through the use of area coordinates and the integral of equation (7.59).

The matrix $[C]$ is identical to that of equation (5.72).

Fig. 7.12 shows the first two eigenmodes of a square cantilevered plate which were calculated using this element.

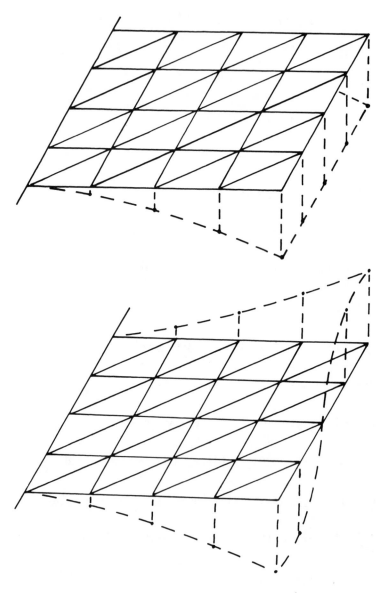

Fig. 7.12 – Modes of vibration of a square cantilevered plate.

7.7.3   Partially conforming plate bending elements

The problem of the plate bending element of Section 7.7.1 is that slope continuity is not satisfied along all three boundaries, as described in Chapter 5. Many methods have been suggested of overcoming this problem, and one of the most important is that due to Clough & Tocher[36], which was based on an idea by Hsieh.

The main problem, however, with the element of Clough & Tocher, was that although it had nine degrees of freedom, it was necessary to set up a 27×27 matrix prior to reducing the stiffness matrix to one of order nine. Furthermore, apart from the fact that it used a large amount of space, it was only suitable for static analysis.

The method suggested in the present text, which was first published in 1973 [48], adopted a similar approach to Clough & Tocher, except that as Guyan reduction[49] was used (Section 7.9), it was applicable to dynamic analysis.

Stiffness and mass matrices are developed by this method for triangular and quadrilateral plate bending elements with corner nodes. The sizes of these matrices are either 9×9 or 12×12, and in developing them it was first necessary to set up 12×12 and 15×15 matrices respectively.

The idea of Hsieh was to assume that the triangular and quadrilateral elements were composed of three or four triangular sub-elements, respectively, as shown in Fig. 7.13 and 7.14. It was then necessary to assume a displacement function for '$\bar{w}$', which gave slope and deflection continuity along the external boundaries of the sub-element, although slope continuity may not have been achieved along the internal boundaries of the sub-elements. A typical sub-element is shown in Fig. 7.15.

Assemblage of the sub-elements, with respect to the x and y axes, followed the usual process, resulting in 12×12 matrices for the triangular element and

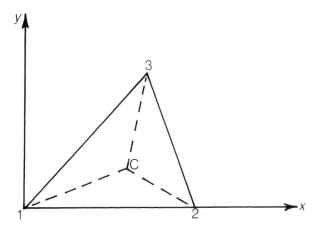

Fig. 7.13 – Triangular element with sub-elements.

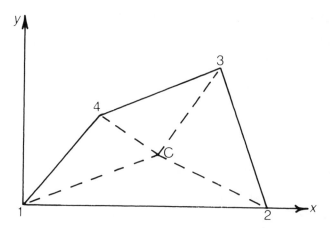

Fig. 7.14 – Quadrilateral element with sub-elements.

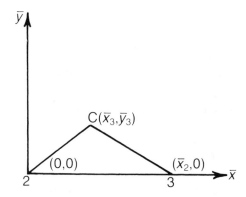

Fig. 7.15 – Triangular sub-element.

15×15 matrices for the quadrilateral element. Elimination of the mid-side node by Guyan reduction resulted in respective 9×9 and 12×12 stiffness and mass matrices for the triangular and quadrilateral elements. The method will now be described.

Consider any triangular sub-element of Fig. (7.13) or (7.14), (say) 2–3–C, as shown in Fig. 7.15. It can readily be shown that an assumed displacement function for the sub-element of Fig. 7.15, which will ensure slope and deflection continuity along its external boundary, 2-3, is

$$\bar{w} = \alpha_1 + \alpha_2 \bar{x} + \alpha_3 \bar{y} + \alpha_4 \bar{x}^2 + \alpha_5 \bar{x}\bar{y} + \alpha_6 \bar{y}^2 + \alpha_7 \bar{x}^3 + \alpha_8 \bar{x}\bar{y}^2 + \alpha_9 \bar{y}^3.$$

$$(7.62)$$

From standard theory it can be shown that

$$[\bar{k}] = [C^{-1}]^{\mathrm{T}} [\tilde{k}] [C^{-1}] \qquad (7.63)$$

and

$$[\bar{m}] = [C^{-1}]^{\mathrm{T}}[\tilde{m}][C^{-1}]$$

where

$$[\tilde{k}] = \int_{\mathrm{vol}} [B_{\bar{x},\bar{y}}]^{\mathrm{T}}[D][B_{\bar{x},\bar{y}}]\mathrm{d}_{\mathrm{vol}} \tag{7.64}$$

and

$$[\tilde{m}] = \rho \int_{\mathrm{vol}} [M_{\bar{x},\bar{y}}]^{\mathrm{T}}[M_{\bar{x},\bar{y}}]\mathrm{d}_{\mathrm{vol}}.$$

Using a process similar to that adopted in sections 5.4 and 7.7.1, it can be seen that $[\bar{k}]$ and $[\bar{m}]$ can be obtained for each sub-element in turn in terms of the $x-y$ coordinate system.

Now for the sub-element, its vector of nodal displacements, namely $\{\bar{u}_i\}$, is defined by

$$\{\bar{u}\} = \{\bar{w}_i \, \bar{\theta}_{xi} \, \bar{\theta}_{yi} \, \bar{w}_j \, \bar{\theta}_{xj} \, \bar{\theta}_{yj} \, \bar{w}_k \, \bar{\theta}_{xk} \, \bar{\theta}_{yk}\} = [C]\{\alpha\}$$

where

$$\bar{\theta}_x = \frac{\partial \bar{w}}{\partial \bar{y}}$$

and

$$\bar{\theta}_y = -\frac{\partial \bar{w}}{\partial \bar{x}}.$$

Hence, $[C]$ can be obtained as in section 5.4 where, for the sub-element of Fig. 7.15, the nodal points, 1, 2, and 3 are the element nodal points 2, 3, and C respectively. For the other sub-elements of Fig. 7.13, the nodal points 1, 2, and 3 correspond to the element nodal points 3–1–C and 1–2–C, respectively.

Now the vector of 'strains' for the sub-element, namely $\{\varepsilon\}$, is given by

$$\{\varepsilon\} = \left[\frac{\partial^2 \bar{w}}{\partial \bar{y}^2} \, -\frac{\partial^2 \bar{w}}{\partial \bar{x}^2} \, \frac{2\partial^2 \bar{w}}{\partial \bar{x}\,\partial \bar{y}} \right]^{\mathrm{T}} = [B_{\bar{x},\bar{y}}]\{\alpha\}$$

so that, $[B_{\bar{x},\bar{y}}]$ can be obtained as in section 5.4. Now,

$$\{\sigma\} = \{M_{\bar{x}} \, M_{\bar{y}} \, M_{\bar{x}\bar{y}}\} = [D]\{\varepsilon\}$$

where $[D]$ is as in equation (5.73). Hence, $[\tilde{k}]$ and $[\tilde{m}]$ can be obtained by the method adopted in section 5.4, where the integrals can be obtained from similar expressions given in Chapter 5 and section 7.7.2

7.7.4 Comparison between theory and experiment

In order to compare the various theories with experiment, the square cantilevered plate of Fig. 7.16 was considered. The experimental observations were due to Barton[50], and the three theories used were as follows:

(a) Non-conforming triangular element[36] of section 7.7.1.
(b) Partially conforming triangular element of section 7.7.3.
(c) Partially conforming quadrilateral of section 7.7.3.

Four different choices of mesh were made for the triangular elements and also for the quadrilateral element, the degree of refinement of the meshes increasing with 'N', and typical meshes for $N = 2$, as shown in Fig. 7.17.

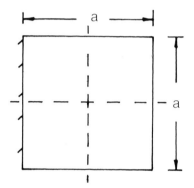

Fig. 7.16 – Square cantilever plate.

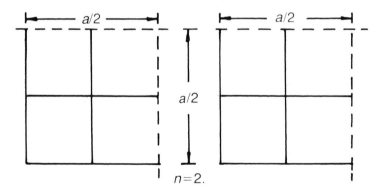

Fig. 7.17 – Typical meshes.

Plots of the natural frequencies for the first three modes, against 'n', are shown in Figs 7.18 to 7.20. From Figs 7.18 to 7.20, it can be seen that the partially conforming elements appear to behave more satisfactorily than the non-conforming triangular element, and also that the quadrilateral element appears to be the best of this family of three elements.

It can be concluded that, although continuity between internal boundaries was not made a prerequisite, application of the resulting elements to a cantilever plate was satisfactory. This was probably because continuity of

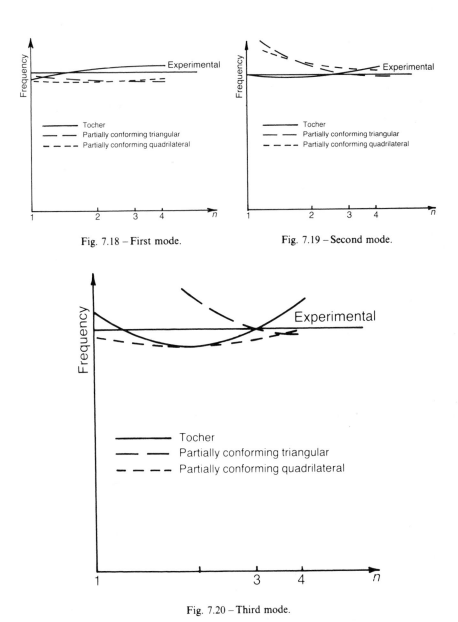

Fig. 7.18 – First mode.   Fig. 7.19 – Second mode.

Fig. 7.20 – Third mode.

slope and deflection was achieved between the elements, and this effect was more significant than the omission of continuity between the internal boundaries of the sub-elements. Although it is likely that these elements are not as sophisticated as some other elements, their simplicity of derivation, together with their compactness, should be attractive. Simplicity and compactness are not often associated with more refined elements.

7.8 DOUBLY CURVED SHELLS

In a manner similar to that adopted for the stiffness matrices of section 5.5, elementary mass matrices in local coordinates can be obtained for doubly curved 'shells', by superimposing the mass matrices determined for in-plane plate elements with those obtained for out-of-plane plate elements, as shown by equation (7.65).

$$[m_s] = \begin{bmatrix} m_{11} & m_{12} & m_{13} \\ m_{21} & m_{22} & m_{23} \\ m_{31} & m_{32} & m_{33} \end{bmatrix} \tag{7.65}$$

where the elements of (7.65) are the following 6×6 sub-matrices:

$$[m_{ij}] = \begin{array}{c} \\ \\ \begin{bmatrix} m_p^{xx} & m_p^{xy} & 0 & 0 & 0 & 0 \\ m_p^{yx} & m_p^{yy} & 0 & 0 & 0 & 0 \\ 0 & 0 & m_b^{zz} & m_b^{zx} & m_b^{zy} & 0 \\ 0 & 0 & m_b^{xz} & m_b^{xx} & m_b^{xy} & 0 \\ 0 & 0 & m_b^{yz} & m_b^{yx} & m_b^{yy} & 0 \\ 0 & 0 & 0 & 0 & 0 & 0 \end{bmatrix} \begin{array}{l} u \\ v \\ w \\ \theta_x \\ \theta_y \\ \theta_z \end{array} \end{array} \tag{7.66}$$

where the subscripts p and b denote in-plane and bending coefficients (see section 5.5).

In global coordinates, the mass matrix becomes

$$[m] = [\Xi]^T [m_s] [\Xi]$$

where

$[\Xi]$ is as in (3.31).

7.8.1

The present author has written two computer programs for analysing doubly curved shells[51], using the triangular plate bending elements described in sections 7.7.1 and 7.7.2, together with the in-plane triangular element of section 5.5. Details of the two programs are as follows:

    (a) FENCT – This program adopts the non-conforming triangular plate bending element of section 7.7.1[36], together with the constant stress in-plane triangular element of Turner et al.[5].

    (b) FEPCT – This program adopts the partially conforming plate bending element of section 7.7.3[48], together with the constant stress in-plane triangular element of Turner et al.

For both programs, as there were six degrees of freedom per mode, it was necessary to adopt the eigenvalue economiser method of Irons[40], which is described in section 7.9. Using this method, it was possible to reduce the sizes of the system matrices by a considerable amount.

Details of a study are given below.

7.8.2 Model wing

This model wing was constructed from bright steel at Portsmouth Polytechnic. It had two kunckles, one near the clamped end and the other along the trailing edge. To assist construction, a small cut was made where the two knuckles met, but prior to any experimental tests being carried out, this cut was brazed and filed smooth.

One end of the model wing was clamped firmly to a vertically moving electromagnetic shaker, as shown in Fig. 7.21. The three lowest resonant frequencies of vibration were detected easily by the naked eye, but stroboscopic light was used to aid observation.

Fig. 7.21 – Model wing with clamping arrangement.

The theoretical analysis was made by the FEPCT program, and the three meshes of Fig. 7.22 were used. From Fig. 7.22, it can be seen that these meshes increased in fineness.

The results are given in Table 7.1, and the first three theoretical eigenmodes for mesh 3 are plotted in Fig. 7.23. The theoretical and experimen-

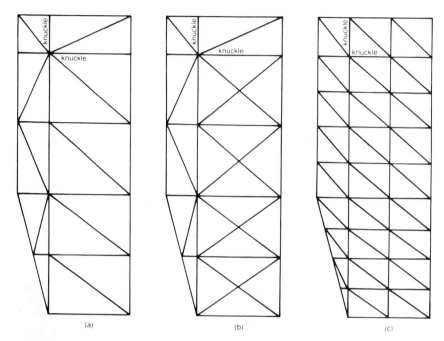

Fig. 7.22(a) – Mesh 1; (b) Mesh 2; (c) Mesh 3.

Table 7.1   Frequencies of model wing (Hz)

Mode	Mesh 1	Mesh 2	Mesh 3	Exptl.	Type
1	66.5	64.1	65.5	49	B
2	190.2	187.6	181.8	172	T
3	376.7	364.8	368.7	308	C

B – bending; T – twisting; C – combined bending and twisting.

tal eigenmodes agreed favourably, and from Table 7.1, it can be seen that the FEPCT element slightly overestimated the experimental results. This may have been because the meshes were not sufficiently refined near the root of the wing, resulting in the mathematical models being too stiff, particularly in bending.

In determining the frequencies for mesh 3, the order of the system matrices was reduced from 186 to 21, using the continuous reduction technique of Irons, which is described in section 7.9.

Another source of error may have been in the assumption for material properties, which were assumed to be

$$E = 2.07 \times 10^{11} \text{ N/m}^2, \quad v = 0.3, \quad \rho = 7830 \text{ kg/m}^3.$$

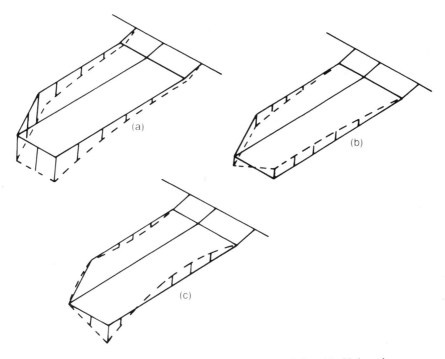

Fig. 7.23(a) – First mode – bending; (b) second mode – twisting; (c) third mode – combined bending and twisting.

One important side effect occurred during the experimental tests, when a crack appeared in the brazing where the two knuckles met. It was decided to determine the three lowest resonant frequencies of vibration with the structure in this condition. These experimental observations revealed the fundamental bending frequency decreased from 49 Hz to 42 Hz, but that the crack had little or no effect on the other two frequencies. It was thought that this was because the fundamental frequency was a pure bending one, whilst the other two frequencies were not.

7.9 REDUCTION OF MASS MATRIX

In a manner similar to that adopted for the stiffness matrix in static analysis, Guyan[49] has shown how it is possible to reduce the size of the mass matrix in dynamic analysis. For this case, it is necessary to ensure that the kinetic energy of the condensed form is the same as that of the uncondensed system.

Suppose that a mass matrix $[M]$ is subdivided into

$$
\begin{bmatrix}
M_{aa} & M_{ab} \\
\hline
M_{ba} & M_{bb}
\end{bmatrix}
$$

so that the dynamical equation becomes,

$$\left\{\begin{matrix} q_a \\ -- \\ 0 \end{matrix}\right\} = \left(\left[\begin{array}{c|c} K_{aa} & K_{ab} \\ \hline K_{ba} & K_{bb} \end{array}\right] - \omega^2 \left[\begin{array}{c|c} M_{aa} & M_{ab} \\ \hline M_{ba} & M_{bb} \end{array}\right]\right) \left\{\begin{matrix} u_a \\ -- \\ u_b \end{matrix}\right\}.$$

If the kinetic energy is to remain the same for both systems, then

$$\tfrac{1}{2}\lfloor u_a{}^T \ u_b{}^T \rfloor \left[\begin{array}{c|c} M_{aa} & M_{ab} \\ \hline M_{ba} & M_{bb} \end{array}\right] \left\{\begin{matrix} u_a \\ u_b \end{matrix}\right\}$$

$$= \tfrac{1}{2}\{u_a\}^T[M_c]\{u_a\}$$

$$= \tfrac{1}{2}\{u_a\}^T[M_{aa}]\{u_a\} + \tfrac{1}{2}\{u_a\}^T[M_{ab}]\{u_b\}$$

$$+ \tfrac{1}{2}\{u_b\}^T[M_{ba}]\{u_a\} + \tfrac{1}{2}\{u_b\}^T[M_{bb}]\{u_b\},$$

but from static considerations (5.88),

$$\{u_b\} = -[K_{bb}]^{-1}[K_{ba}]\{u_a\},$$

therefore

$$\text{K.E.} = \tfrac{1}{2}\{u_a\}^T[M_{aa}]\{u_a\} - \tfrac{1}{2}\{u_a\}^T[M_{ab}][K_{bb}]^{-1}[K_{ba}]\{u_a\}$$

$$- \tfrac{1}{2}\{u_a\}^T([K_{bb}]^{-1}[K_{ba}])^T[M_{ba}]\{u_a\}$$

$$+ \tfrac{1}{2}\{u_a\}^T([K_{bb}]^{-1}[K_{ba}])^T[M_{bb}][K_{bb}]^{-1}[K_{ba}]\{u_a\},$$

that is,

$$[M_c] = [M_{aa}] - [M_{ab}][K_{bb}]^{-1}[K_{ba}] - ([K_{bb}]^{-1}[K_{ba}])^T$$

$$\times ([M_{ba}] - [M_{bb}][K_{bb}]^{-1}[K_{ba}]), \qquad (7.67)$$

where

$$[M_c] = \text{condensed mass matrix.}$$

Unlike the static case, some loss of precision will occur, owing to the fact that a static relationship has been used to link the dynamic displacements $\{u_b\}$ and $\{u_a\}$. If, however, the mass matrix is carefully subdivided so that $\{u_b\}$ is insignificant compared with $\{u_a\}$, then these errors will be small.

7.9.1 Continuous reduction technique

Although the method of Guyan is a powerful one, it is not suitable for large problems, because $[K_{bb}]$ has to be inverted. Irons[40] has overcome this difficulty with a continuous reduction technique, which, if properly applied, can reduce the size of the problem by a factor of 20 or more with little loss of precision.

The process of Irons is to define displacements as 'slaves' and 'masters', and to reduce the size of the problem, so that only the equations corresponding to 'master' displacements remain. 'Master' displacements are of course much more important than 'slave' displacements, and the problem of distinguishing between these is partly a matter of judgement and partly one of experience. For example, with a simple cantilever subdivided into several elements, obvious 'slaves' would be rotational displacements and also lateral deflections near the root of the cantilever. A similar analogy could be applied to flat plates, but it should be remembered that although this process would tend to satisfactorily diagnose the lower modes of vibration, some higher modes are likely to be missed out altogether.

Irons's continuous reduction technique is given by equations (7.68), where it can be seen that it really is an extension of Guyan's method.

When a displacement $u_s{}^o$ becomes a 'slave', the row and column corresponding to 's' are deleted from $[K^o]$ and $[M^o]$, and their 'ij' terms become

$$K_{ij}^* = K_{ij} - K_{is}(K_{sj}/K_{ss})$$
$$M_{ij}^* = M_{ij} - M_{is}(K_{sj}/K_{ss}) - M_{sj}(K_{is}/K_{ss}) + M_{ss}(K_{sj}/K_{ss})(K_{is}/K_{ss})$$
$$(7.68)$$

The main advantage of the method is that, once the elements of the stiffness and mass matrix corresponding to a slave displacement are complete, it is possible to eliminate it, even though other slaves and masters are incomplete. It should, however, be ensured that before eliminating a slave, that 'lower' slaves have been previously eliminated, but providing this has been done, the space available after the elimination can be used for storing 'higher' slaves and masters. Thus it can be seen that the process is a continuous reduction technique, and very useful for large problems. It should be noted that it is possible to use this method for static problems, but for these cases the 'slaves' should be taken at the nodes with zero loads.

7.10   UNITS FOR VIBRATION ANALYSIS

Convenient units may be taken as follows:

SI
Mass (kg); Force (N); Time (s); Density (kg/m^3).

Imperial
Mass (lbf.s^2/in); Force (lbf); Time (s); Density (lbf.s^2/in^4).

7.11   MICROCOMPUTER PROGRAMS

Computer programs for the free vibration of plane and space pin-jointed trusses, continuous beams, and rigid-jointed plane frames, are given in reference [15].

Using the programs 'VIBPTR' and 'VIBSPTR' of reference [15], the plane pin-jointed trusses of Example 3.2, together with the space truss of Example 7.3, were analysed on an Apple II microcomputer.

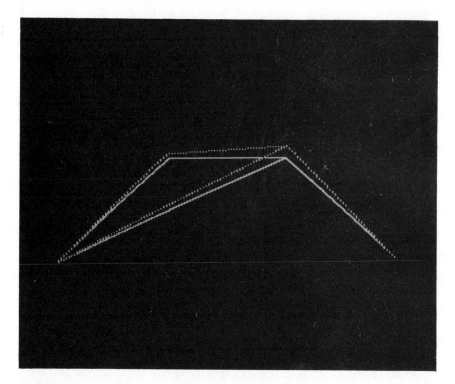

(a) 1st eigenmode.

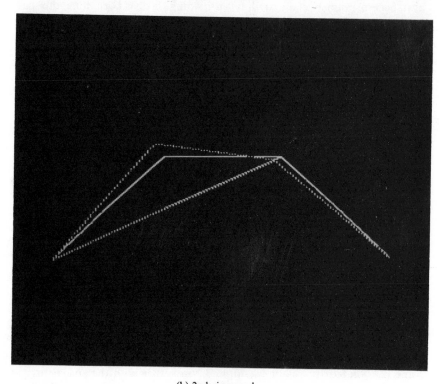

(b) 2nd eigenmode.

Fig. 7.24 – Eigenmodes of a plane truss (Example 3.2).

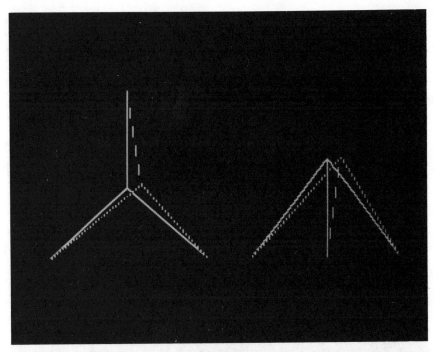

(a) 1st eigenmode.

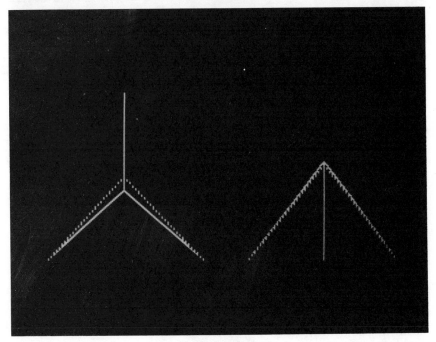

(b) 2nd eigenmode.

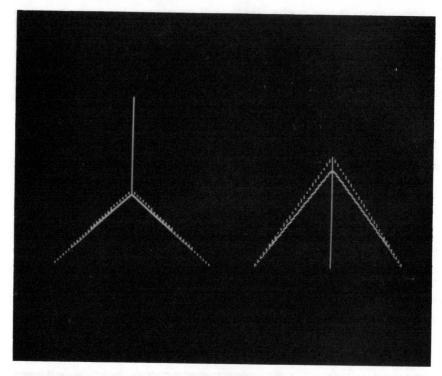

(c) 3rd eigenmode.

Fig. 7.25 – Eigenmodes of 2 space truss (Example 7.3).

Figs 7.24 and 7.25 show the eigenmodes of these trusses, where for the second, the plan is drawn on the left of the screen and the front elevation on the right.

EXAMPLES FOR PRACTICE

1. Calculàte the two lowest axial frequencies of vibration, together with their eigenmodes, for a uniform rod of length 'l', fixed at node 1 and free at node 2, using two equal length elements.

   Answer

   $$\omega_1 = 1.61 \sqrt{(E/\rho/l}; \lfloor u_2 \quad u_3 \rfloor = [0.707 \quad 1\rfloor$$
   $$\omega_2 = 5.63 \sqrt{(E/\rho/l}; \lfloor u_2 \quad u_3 \rfloor = [-0.707 \quad 1\rfloor.$$

   Exact answer: $\omega_1 = 1.571 \sqrt{(E/\rho/l}.$

2. Calculate the two lowest natural axial frequencies of vibration, together with their eigenmodes, for an unconstrained rod, using a one element model.

   Answer

   $$\omega_1 = 0; \lfloor u_1 \quad u_2 \rfloor = \lfloor 1 \quad 1 \rfloor - \text{Rigid body mode}$$
   $$\omega_2 = 3.46 \sqrt{(E/\rho}/l; \lfloor u_1 \quad u_2 \rfloor = \lfloor -1 \quad 1 \rfloor.$$

   Exact answers: $\omega_1 = 0; \omega_2 = \pi\sqrt{(E/\rho}/l.$

3. Determine the three lowest natural frequencies of vibration for the previous problem, using two equal length elements.

   Answer

   $$\omega_1 = 0; \lfloor u_1 \quad u_2 \quad u_3 \rfloor = \lfloor 1 \quad 1 \quad 1 \rfloor - \text{Rigid body mode}$$
   $$\omega_2 = 3.46 \sqrt{(E/\rho}/l; \lfloor u_1 \quad u_2 \quad u_3 \rfloor = \lfloor -1 \quad 0 \quad 1 \rfloor$$
   $$\omega_3 = 6.93 \sqrt{(E/\rho}/l; \lfloor u_1 \quad u_2 \quad u_3 \rfloor = \lfloor 1 \quad -1 \quad 1 \rfloor.$$

   Exact answers: $\omega_1 = 0; \omega_2 = \pi\sqrt{(E/\rho}/l; \omega_3 = 2\omega_2.$

4. Using the method of minimum potential, derive the mass matrix for the plane uniform beam of Fig. 7.2.

   Answer

$$\frac{\rho A l}{420} \begin{array}{cccc} v_1 & \theta_1 & v_2 & \theta_2 \\ \begin{bmatrix} 156 & & & \\ -22l & 4l^2 & & \\ 54 & -13l & 156 & \\ 13l & -3l^2 & 22l & 4l^2 \end{bmatrix} & \begin{array}{c} v_1 \\ \theta_1 \\ v_2 \\ \theta_2 \end{array} \end{array}$$

5. Calculate the lowest natural frequency of vibration for a plane uniform beam, pinned at both ends and using a one element model.

   Answer

   $$10.95 \sqrt{(EI/\rho A)}/l^2.$$

   Exact answer: $9.85 \sqrt{(EI/\rho A)}/l^2.$

6. Calculate the lowest natural frequency of vibration for a plane uniform beam, clamped at both ends, using two equal length elements.

   Answer

   $$\omega = 22.76 \sqrt{(EI/\rho A)}/l^2.$$

   Exact answer: $= 22.2 \sqrt{(EI/\rho A)}/l^2.$

7. Calculate the two lowest natural frequencies of vibration for the plane pin-jointed truss shown in Fig. 7.26, given the following:

$A = 0.001 \text{ m}^2$, $E = 2 \times 10^{11} \text{ N/m}^2$, $\rho = 7860 \text{ kg/m}^3$.

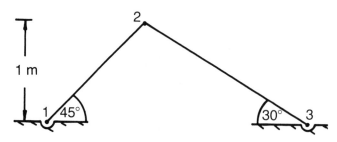

Fig. 7.26.

$n_1 = 487 \text{ Hz}; \lfloor u_2{}^\circ \quad v_2{}^\circ \rfloor = \lfloor -0.44 \quad 1 \rfloor$

$n_2 = 668 \text{ Hz}; \lfloor u_2{}^\circ \quad v_2{}^\circ \rfloor = \lfloor 1 \quad 0.44 \rfloor$.

8. Calculate the two lowest natural frequencies of vibration for the plane pin-jointed truss of Problem 7, assuming that there is an additional mass of 20 kg at node 2.

$n_1 = 270.6 \text{ Hz}; \lfloor u_2{}^\circ \quad v_2{}^\circ \rfloor = \lfloor -0.44 \quad 1 \rfloor$

$n_2 = 371.4 \text{ Hz}; \lfloor u_2{}^\circ \quad v_2{}^\circ \rfloor = \lfloor 1 \quad 0.44 \rfloor$.

9. Calculate the two lowest natural frequencies of vibration of the plane pin-jointed trusses of Problems (1a) and (1c) from 'Examples for Practice', Chapter 3, given the following:

$A = 0.001 \text{ m}^2$, $\rho = 7860 \text{ kg/m}^3$, $E = 2 \times 10^{11} \text{ N/m}^2$.

Answer

1(a)

$n_1 = 555.2 \text{ Hz}; \lfloor u_4{}^\circ \quad v_4{}^\circ \rfloor = \lfloor 1 \quad 0.18 \rfloor$

$n_2 = 811.6 \text{ Hz}; \lfloor u_4{}^\circ \quad v_4{}^\circ \rfloor = \lfloor -0.18 \quad 1 \rfloor$

1(c)

$n_1 = 359.2 \text{ Hz}; \lfloor u_4{}^\circ \quad v_4{}^\circ \rfloor = \lfloor -0.27 \quad 1 \rfloor$

$n_2 = 447.97 \text{ Hz}; \lfloor u_4{}^\circ \quad v_4{}^\circ \rfloor = \lfloor 1 \quad 0.28 \rfloor$

10. Calculate the three lowest natural frequencies of vibration of the pin-jointed space trusses of Problems (2a) and (2b), from 'Examples for practice', Chapter 3, given the following

$A = 0.001 \text{ m}^2$, $\rho = 7860 \text{ kg/m}^3$, $E = 2 \times 10^{11} \text{ N/m}^2$

Ans.

2(a) $n_1 = 95.7$ Hz; $\lfloor u_5^\circ \quad v_5^\circ \quad w_5^\circ \rfloor = \lfloor 1 \quad 0.38 \quad 0.26 \rfloor$

$\quad\quad n_2 = 101.5$ Hz; $\lfloor u_5^\circ \quad v_5^\circ \quad w_5^\circ \rfloor = \lfloor 0 \quad 1 \quad 0 \rfloor$

$\quad\quad n_3 = 161.3$ Hz; $\lfloor u_5^\circ \quad v_5^\circ \quad w_5^\circ \rfloor = \lfloor -0.26 \quad 0.02 \quad 1 \rfloor$

2(b) $n_1 = 46.4$ Hz; $\lfloor u_6^\circ \quad v_6^\circ \quad w_6^\circ \rfloor = \lfloor 0.19 \quad 1 \quad 0 \rfloor$

$\quad\quad n_2 = 66.4$ Hz; $\lfloor u_6^\circ \quad v_6^\circ \quad w_6^\circ \rfloor = \lfloor 1 \quad 0 \quad -0.21 \rfloor$

$\quad\quad n_3 = 109.2$ Hz; $\lfloor u_6^\circ \quad v_6^\circ \quad w_6^\circ \rfloor = \lfloor 0.21 \quad 0 \quad 1 \rfloor.$

Chapter 8

Grillages

This chapter is concerned with the static and vibration analysis of orthogonal and skew grillages, and a number of such structures are analysed by computer in this chapter.

Grillages or grids are often used to cover wide areas, where the supports can, in general, only be placed around the periphery of the structure and at a few other discrete points. For such structures (see Fig. 8.1), the stiffening effect of cross-members is used to strengthen the structure so that the lateral loads can be resisted in bending and torsion by the members of the grid in both directions.

Grillage structures are of much interest in many branches of engineering, including the aircraft, shipbuilding, and the mechanical and civil engineering industries. These structures take many various forms, including hatch covers, decks, platforms, floors, etc.

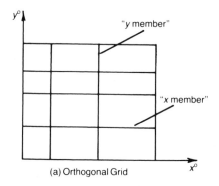

(a) Orthogonal Grid

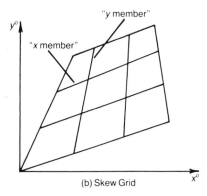

(b) Skew Grid

Fig. 8.1 – Orthogonal and skew grids.

Much of the work that has been done on grillage structures has been based on analytical techniques [52 to 55], and these theories have usually considered the static analysis of orthogonally stiffened rectangular grillages under uniform pressure. Whereas these theories are very useful for a large number of problems, they break down for grillages of complex shape with complex boundary conditions and load, and where the members of the grillage are made from different sections.

These deficiencies have prompted the introduction of numerical methods and, in particular, the finite element method, which is now described.

8.1 ELEMENTAL STIFFNESS AND MASS MATRICES

The element in local coordinates is shown in Fig. 8.2, and the elemental stiffness matrix, in local coordinates, which is a combination of a beam element (Eq. (4.13)) and a torque bar (Eq. (4.39)), can readily be shown to take the form of equation (8.1).

$$[k] = \begin{array}{c} \begin{array}{cccccc} w_1 \quad & \theta_{x1} \quad & \theta_{y1} \quad & w_2 \quad & \theta_{x2} \quad & \theta_{y2} \end{array} \\ \left[\begin{array}{cccccc} 12EI/l^3 & & & & & \\ 0 & GJ/l & & & & \\ -6EI/l^2 & 0 & 4EI/l & & & \\ \hline -12EI/l^3 & 0 & 6EI/l^2 & 12EI/l^3 & & \\ 0 & -GJ/l & 0 & 0 & GJ/l & \\ -6EI/l^2 & 0 & 2EI/l & 6EI/l^2 & 0 & 4EI/l \end{array} \right] \begin{array}{c} w_1 \\ \theta_{x1} \\ \theta_{y1} \\ w_2 \\ \theta_{x2} \\ \theta_{y2} \end{array} \end{array}$$

$$\text{(8.1)}$$

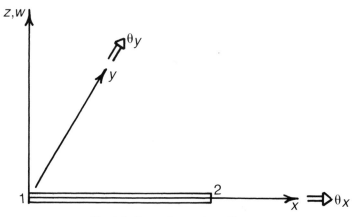

Fig. 8.2 – Beam element for grillage.

Similarly, the elemental mass matrix for the grillage element, in local coordinates, is a simple combination of the mass matrices for a beam element (Eq. (7.41)) and a torque bar (Eq. (7.50)), as shown in (8.2).

$$
[m] = \rho AL
\begin{bmatrix}
\dfrac{13}{35} + \dfrac{6I}{5Al^2} & & & \\[2ex]
0 & \dfrac{I_p}{3A} & & \\[2ex]
\dfrac{-11l}{210} - \dfrac{I}{10Al} & 0 & \dfrac{l^2}{105} + \dfrac{2I}{15A} & \\[2ex]
\hline
\dfrac{9}{70} - \dfrac{6I}{5Al^2} & 0 & \dfrac{-13l}{420} + \dfrac{I}{10Al} & \\[2ex]
0 & \dfrac{I_p}{6A} & 0 & \\[2ex]
\dfrac{13l}{420} - \dfrac{I}{10Al} & 0 & \dfrac{l^2}{140} - \dfrac{I}{30A} &
\end{bmatrix}
\begin{matrix} w_1 \\ \theta_{x1} \\ \theta_{y1} \end{matrix}
$$

$$
\begin{array}{cccc}
 & w_2 & \theta_{x2} & \theta_{y2} \\
\end{array}
$$

$$
\left[
\begin{array}{ccc}
 & & \\
 & & \\
 & & \\
\hline
\dfrac{13}{35}+\dfrac{6I}{5Al^2} & & \\
0 & \dfrac{I_p}{3A} & \\
\dfrac{111}{210}+\dfrac{I}{10Al} & 0 & \dfrac{l^2}{105}+\dfrac{2I}{15A}
\end{array}
\right]
\begin{array}{l}
w_1 \\
\theta_{x1} \\
\theta_{y1} \\
\\
w_2 \\
\\
\theta_{x2} \\
\\
\theta_{y2}
\end{array}
\qquad (8.2)
$$

where

l = length of element
I = 2nd moment of area about the x–y plane
J = torsional constant
I_p = 2nd polar moment of area about the 'x' axis
A = cross-sectional area
E = elastic modulus
G = rigidity modulus
ρ = density.

In practice, however, it is necessary to obtain the elemental stiffness and mass matrices in the global coordinates of Fig. 8.3. This is because some of the grillage members point in directions other than that of Fig. 8.2.

Transformation of the stiffness and mass matrices of (8.1) and (8.2) into global axes is achieved as follows:

$[k^\circ]$ = elemental stiffness matrix in global axes
$\quad = [\Xi]^{\mathrm{T}}[k][\Xi]$

and

$[m^\circ]$ = elemental mass matrix in global axes
$\quad = [\Xi]^{\mathrm{T}}[m][\Xi],$

where

$$
[\Xi] = \left[
\begin{array}{c|c}
\zeta & 0_3 \\
\hline
0_3 & \zeta
\end{array}
\right]
$$

$$[\zeta] = \begin{bmatrix} 1 & 0 & 0 \\ 0 & c & s \\ 0 & -s & c \end{bmatrix}$$

O_3 = a null matrix of order 3.

$c = \cos \alpha$

$s = \sin \alpha$

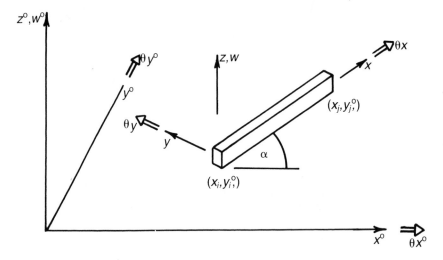

Fig. 8.3 – Grillage element in global coordinates.

The relationship between local and global displacements is

$$\{u_i\} = [\Xi] \{u_i{}^o\}$$

where

$$\{u_i\} = \lfloor w_i \quad \theta_{xi} \quad \theta_{yi} \rfloor^{\mathrm{T}}$$

and

$$\{u_i{}^o\} = \lfloor w_i{}^o \quad \theta_{xi}{}^o \quad \theta_{yi}{}^o \rfloor^{\mathrm{T}}.$$

The displacement w_i is equal to $w_i{}^o$, as the z axis is parallel to the z^o axis.

For the static analysis of uniform section beams and torque bars, the following relationships are well-known:

$$T = GJ \frac{d\phi}{dx}$$

$$M = EI \frac{d^2w}{dx^2} \tag{8.3}$$

where

T = torque

M = bending moment

$\dfrac{d\phi}{dx}$ = angle of twist/unit length.

Now, from equation (5.63),

$$w = \left[(1 - 3\xi^2 + 2\xi^3) \, l(-\xi + 2\xi^2 - \xi^3) \, (3\xi^2 - 2\xi^3) \, l(\xi^2 - \xi^3) \right] \begin{Bmatrix} w_1 \\ \theta_{y1} \\ w_2 \\ \theta_{y2} \end{Bmatrix},$$

therefore

$$\frac{d^2 w}{dx^2} = \frac{d^2 w}{l^2 d\xi^2} = \frac{(6 - 12\xi)}{l^2} w_1 + \frac{(-4 + 6\xi)}{l} \theta_{y1}$$

$$- \frac{(6 - 12\xi)}{l^2} w_2 + \frac{(-2 + 6\xi)}{l} \theta_{y2} \tag{8.4}$$

and from section 7.4.4,

$$\phi = (1 - x/l)\phi_1 + (x/l)\phi_2$$

so that

$$\frac{d\phi}{dx} = \frac{d\phi}{l d\xi} = -\phi_1/l + \phi_2/l, \tag{8.5}$$

where

ϕ_1 = angle of twist at node 1 = θ_{x1}

ϕ_2 = angle of twist at node 2 = θ_{x2}

$\xi = x/l$.

Substituting (8.4) and (8.5) into (8.3) and rewriting in matrix form, the following is obtained:

$$\begin{Bmatrix} T \\ M \end{Bmatrix} = [D][B]\{u_i\}, \tag{8.6}$$

where

$$[D] = \begin{bmatrix} GJ & 0 \\ 0 & EI \end{bmatrix}$$

$$[B] = \begin{bmatrix} 0 & -1/l & 0 & 0 & 1/l & 0 \\ \dfrac{(6-12\xi)}{l^2} & 0 & \dfrac{(-4+6\xi)}{l} & -\dfrac{(6-12\xi)}{l^2} & 0 & \dfrac{(-2+6\xi)}{l} \end{bmatrix}.$$

8.2   STATIC ANALYSIS OF GRILLAGES BY COMPUTER

The program 'GRILLAGE' of reference [15] can analyse orthogonal or skew grids of complex shape and boundary condition. It can easily be modified to cater for more sophisticated or less sophisticated problems.

8.2.1   Example 8.1

Determine the deflections, torques, and bending moments for the uniformly spaced simply-supported grillage shown in Fig. 8.4. This grillage is the same as

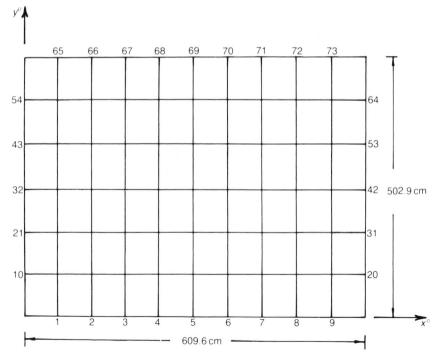

Fig. 8.4 – Simply supported grillage.

that used as an example in reference [55]. It was subjected to an upward uniform pressure of 6.895 kN/m², and this was assumed to be equivalent to uniformly distributed loads of 2.79 kN/m in both x° and y° directions.

The details of the grillage are as follows:

$$E = 6.89 \times 10^{10} \text{ N/m}^2$$
$$G = 2.65 \times 10^{10} \text{ N/m}^2$$
$$I_x = 1.349 \times 10^{-4} \text{ m}^4$$
$$I_y = 4.495 \times 10^{-5} \text{ m}^4$$
$$I_{px} = J_x = 1.199 \times 10^{-6} \text{ m}^4 \text{ (assumed)}$$
$$I_{py} = J_y = 6.04 \times 10^{-7} \text{ m}^4 \text{ (assumed)}.$$

The bending moment distributions for some of the members are given in Figs 8.5 and 8.6, where they are compared with the Vedeler [52] solution. It can be seen from Figs 8.5 and 8.6 that there is good agreement between the matrix solution and the Vedeler solution.

Once again, the bending moment predictions by the matrix solution prove to be larger than the analytical solution, and this is believed to be due to the loading assumptions. The Vedeler solution is based on orthotropic plate theory and assumes that some of the loading is transmitted to the boundary via the plating, whereas the matrix solution assumes that all the loading is transmitted to the boundary via the grid.

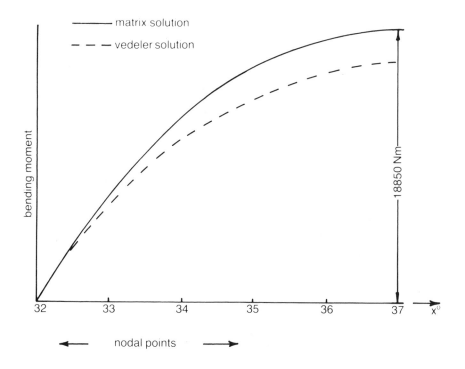

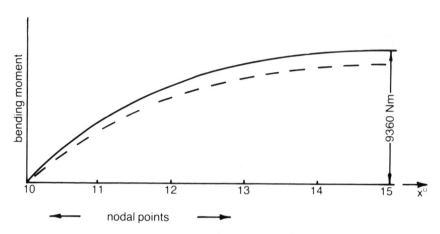

Fig. 8.5 – Bending moment diagrams for some members in "*x*" direction.

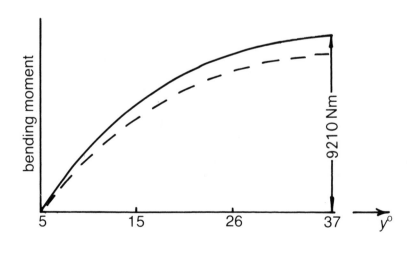

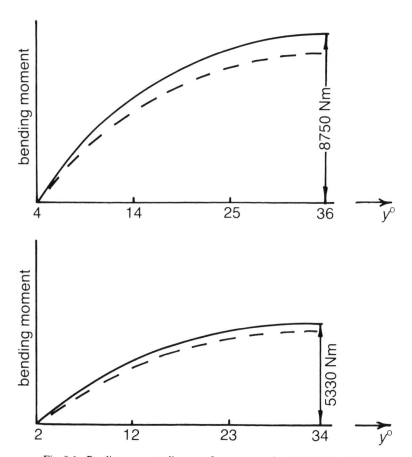

Fig. 8.6 – Bending moment diagrams for some members in "y" direction.

8.2.2 Example 8.2

Determine deflections, torques and bending moments for the ship grillage shown in Fig. 8.7 using "GRILLAGE" of reference [15]. The x^o direction members are subjected to a downward uniformly distributed load of 500 N/m, and there are two downward concentrated loads of 4000 N and 3000 N acting at nodes 15 and 20. All edges may be assumed to be simply-supported.

The members may be assumed to have the following properties:

$$I_x = 8.6 \times 10^{-7} \text{ m}^4$$
$$I_y = 3.6 \times 10^{-6} \text{ m}^4$$
$$I_{px} = J_x = 9 \times 10^{-8} \text{ m}^4 \text{ (assumed)}$$
$$I_{py} = J_y = 2.2 \times 10^{-7} \text{ m}^4 \text{ (assumed)}$$
$$E = 2 \times 10^{11} \text{ N/m}^2$$
$$G = 7.69 \times 10^{10} \text{ N/m}^2.$$

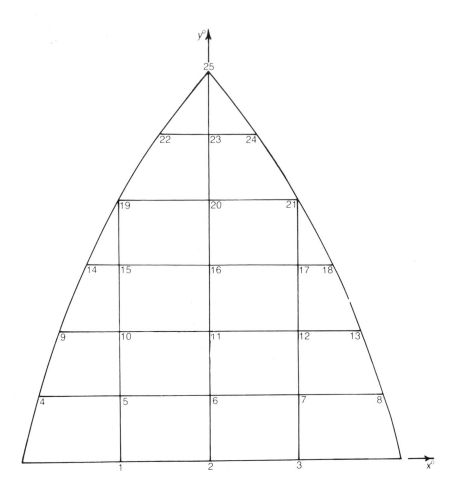

Fig. 8.7 – Ship grillage.

The bending moment distributions for some of the members are shown in Fig. 8.8. The apparent discontinuities in the bending moment diagrams at some of the joints is because of the torsional resistance of other members lying perpendicular to these members at these joints. Hence, balancing of the joints is in fact achieved if consideration is made of couples causing torque in addition to couples causing bending.

The change in sign of bending moment for the central girder at node 23 is due to the comparatively large stiffness of the x^o direction member 22–23–24. This member is particularly stiff because it is very short. In fact, it appears to act in a manner similar to that of an 'elastic prop'.

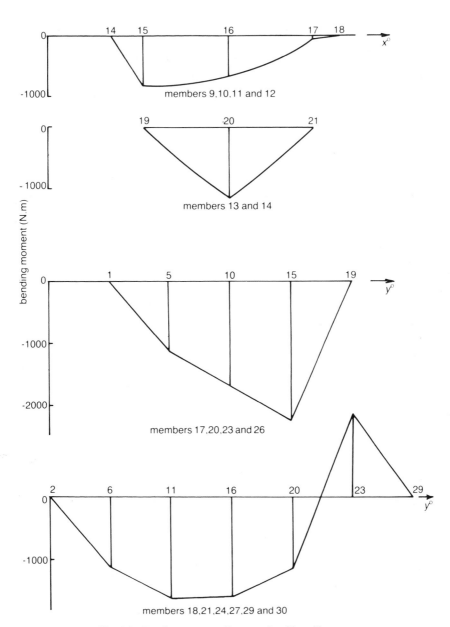

Fig. 8.8 – Bending moment diagrams for ship grillage.

8.3 PROGRAM FOR THE VIBRATION OF FLAT GRILLAGES

This program, which is written for a mainframe, can analyse orthogonal or skew grids of complex shape and boundary condition. It determines natural frequencies and eigenmodes, and the eigenvalue procedure is based on the power method, together with Aitken's acceleration [22].

A precision factor 'D' is set to 0.001, which is equivalent to 0.1 % precision. This can be increased or decreased, but if it is decreased the solution time might increase, and if 'D' is made too small, it is possible that the precision of the computer may be exceeded. A value of 'D' of 0.0001 is equivalent to a precision of 0.01 %, and one of 0.01 to 1 %.

8.3.1   Example 8.3

Determine the natural frequencies and the corresponding eigenmodes for the skew grid in Fig. 8.9. This grid is the same as that adopted by Venancio-Filho &

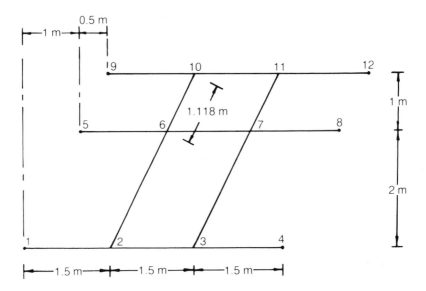

Fig. 8.9 – Skew grillage.

Iguti [56]. It is simply-supported at nodes 1, 4, 5, 8, 9, and 12, and its members have the following properties:

$$A = 0.004 \text{ m}^2,\ I_x = I_y = 1.25 \times 10^{-5} \text{ m}^4$$
$$I_p = J = 2.5 \times 10^{-5} \text{ m}^4 \text{ (assumed)}$$
$$E = 2 \times 10^{11} \text{ N/m}^2,\ G = 7.69 \times 10^{10} \text{ N/m}^2$$
$$\rho = 7860 \text{ kg/m}^3.$$

Results

The frequencies are compared in Table 8.1 with the predictions of Venancio-Filho & Iguti, and the eigenmodes are plotted in Fig. 8.10.

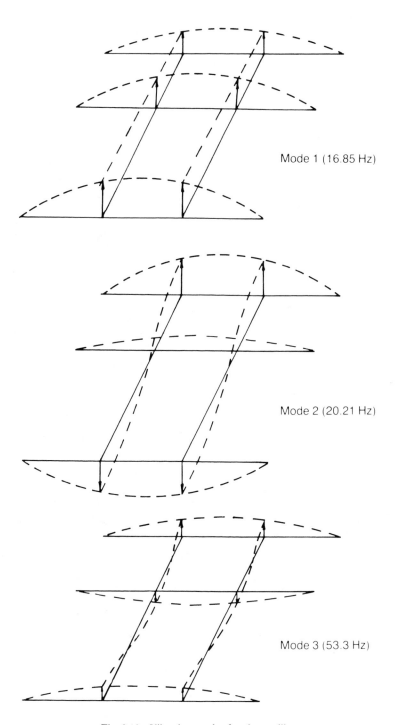

Fig. 8.10 – Vibration modes for skew grillage.

Table 8.1. Frequencies (Hz) for skew grid

Frequencies	Matrix solution	Venancio-Filho & Iguti
n_1	16.85	16.93
n_2	20.21	20.63
n_3	53.30	51.61

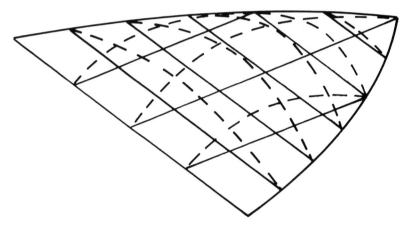

First mode

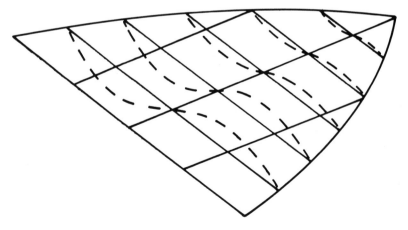

Second mode

Fig. 8.11 – Eigenmodes for ship grillage.

From Table 8.1 it can be seen that the matrix solution compares most favourably with the frequency predictions of Venancio-Filho & Iguti.

8.3.2   Example 8.4

Determine the first three frequencies and the corresponding eigenmodes for the ship grillage analysed in Example 8.2. The following additional data are applicable to the members of the grillage:

$$A_x = 0.0012 \text{ m}^2$$
$$A_y = 0.0032 \text{ m}^2$$
$$\rho = 7860 \text{ kg/m}^3.$$

The eigenmodes corresponding to the first two frequencies are shown in Figure 8.11. The first eigenmode appears to be symmetrical and consists of a half-wave in both directions, and the second eigenmode appears to be unsymmetrical and consists of two half-waves in the x^o direction. These eigenmodes appear to be much as expected.

The close correlation between the matrix solution and other well-known solutions for both static and vibration analysis is very encouraging.

One advantage, however, of the matrix solutions over many other solutions is that they can cater for various combinations of simply-supported and clamped boundary, whether these are internal or external. Other advantages of the matrix solutions are that the boundary shapes can be straight, curved or irregular, and for static analysis, the loads can consist of various combinations of distributed and concentrated load acting upwards and downwards.

Chapter 9

Non-linear structural mechanics

So far, we have considered only the linear elastic theory of structures. However, it will now be shown how this theory can be extended to cover some problems in non-linear structural mechanics.

Non-linear structural analysis consists of the following three types:

(a) *Geometrical non-linearity*, where although the structure is still elastic, the effects of large deflections cause the geometry of the structure to change, so that linear elastic theory breaks down. Typical problems that lie in this category are the elastic instability of structures, such as in the Euler buckling of struts and also the large deflection analysis of plates. For the latter, the geometry of the plate changes, so that membrane effects become significant.

In general, it can be said that for geometrical non-linearity, an axially applied compressive force in a member decreases its bending stiffness, but an axially applied tensile force increases its bending stiffness.

(b) *Material non-linearity*, where the material undergoes plastic deformation.

(c) *Combined geometrical and material non-linearity* A typical case of this is the inelastic instability of struts, such as those analysed by the Rankine–Gordon formula, where initial imperfections are important. There are many structures which should theoretically fail through elastic instability but, because of initial imperfections, fail at a much lower load than the predictions based on elastic theory. Such structures are said to suffer *elastic knockdown*.

9.1 COMBINED GEOMETRICAL AND MATERIAL NON-LINEARITY

The difficulty in applying the stiffness analysis to problems involving instability is that, any compressive axial load on a structural member will tend to decrease its bending stiffness and vice versa for a tensile axial load. Thus, if a structural member is subjected to compressive axial loading, buckling will occur when the bending stiffness is so decreased, that any small bending caused by the offset of the load will be larger than the bending resistance of the structure. Such analysis is further complicated by the effects of material non-linearity where the stresses in certain parts of the structure have exceeded the limit of proportionality.

This behaviour can be readily observed by an examination of the instability of axially-loaded struts. In the matrix method of analysing the instability of such structures, it is necessary to consider these effects by adding to the stiffness matrix $[K]$, an initial stress-stiffness matrix $[K_G]$. This latter matrix usually considers the change of stiffness due to any internal forces in the structure, and makes the linear load–displacement relationship of equation (2.6) inapplicable.

These non-linear terms, together with the material non-linearity effects on $[k]$, modify the resulting stiffness matrix to

$$[k] = [k_o] + [k_G]$$

where

$[k_o] =$ a stiffness matrix. (If the material properties of the element are constant, then this matrix is the usual stiffness matrix.)

$[k_G] =$ the geometrical stiffness matrix which depends on the internal forces in the element. This matrix is also known as the *initial stress-stiffness matrix*.

The corresponding elemental stiffness–displacement relationship is,

$$\{p\} = ([k_o] + [k_G])\{u_i\}. \tag{9.1}$$

Similarly, the total stiffness matrix in global coordinates is given by

$$[K^o] = [K_o^o] + (K_G^o),$$

and the load–displacement relationship is

$$\{q^o\} = ([K_o^o] + [K_G^o])\{u_i\} \tag{9.2}$$

where

$[K_o^o] = \Sigma[k_o^o] =$ system stiffness matrix in global coordinates

$[K_G^o] = \Sigma[k_G^o] =$ system geometrical stiffness matrix coordinates.

The process, therefore adopted, is an incremental one, where the load is increased in small steps and the effects of geometrical and material non-linearity are considered for each step and summed together to give the overall effect at the end of each step.

Turner *et al.* [57] were responsible for the incremental step procedure, and further work was done by Turner *et al.* [58], Martin [59], and Argyris *et al.* [60].

Table 9.1 summarises the incremental step method of Turner *et al.* In step 1, if the initial stresses before loading are zero, then $[K_G]$ is null, so that on application of an incremental load $\{\delta q_1{}^\circ\}$, the displacements $\{\delta u_1{}^\circ\}$ due to this load are calculated and set equal to $\{u_1{}^\circ\}$.

Table 9.1. Incremental step procedure for non-linear analysis

Step	$\{\delta q^\circ\}$	Stiffness matrix	$\{\delta u^\circ\}$	Displacements
1	$\{\delta q_1{}^\circ\}$	$[K_o{}^\circ(0)] + [K_G{}^\circ(0)]$	$\{\delta u_1{}^\circ\}$	$\{u_1{}^\circ\} = \{\delta u_1{}^\circ\}$
2	$\{\delta q_2{}^\circ\}$	$[K_o{}^\circ(u_1{}^\circ)] + [K_G{}^\circ(u_1{}^\circ)]$	$\{\delta u_2{}^\circ\}$	$\{u_2{}^\circ\} = \{u_1{}^\circ\} + \{\delta u_2{}^\circ\}$
3	$\{\delta q_3{}^\circ\}$	$[K_o{}^\circ(u_2{}^\circ)] + [K_G{}^\circ(u_2{}^\circ)]$	$\{\delta u_3{}^\circ\}$	$\{u_3{}^\circ\} = \{u_2{}^\circ\} + \{\delta u_3{}^\circ\}$
.	.	.	.	.
.	.	.	.	.
n	$\{\delta q_n{}^\circ\}$	$[K_o{}^\circ(u^\circ{}_{n-1})] + [K_G{}^\circ(u^\circ{}_{n-1})]$	$\{\delta u_n{}^\circ\}$	$\{u_n{}^\circ\} = \{u^\circ{}_{n-1}\} + \{\delta u_n{}^\circ\}$
Σ	$\{q_n{}^\circ\}$		$\{u_n{}^\circ\}$	

The geometrical non-linear effects of $\{u_1{}^\circ\}$ are then determined for $[K_G{}^\circ]$ and the effects of any material non-linearity are then allowed for in $[K_o{}^\circ]$. These two matrices are then added together in step 2 to give the new stiffness matrix. The structure is now subjected to another vector of incremental loads $\{\delta q_2{}^\circ\}$, and the corresponding vector of displacements $\{\delta u_2{}^\circ\}$ are calculated and added to $\{u_1{}^\circ\}$ to give $\{u_2{}^\circ\}$.

The geometrical non-linear effects of $\{u_2{}^\circ\}$ are then determined for $[K_G{}^\circ\}$, and any material non-linearity is allowed for in $[K_o{}^\circ]$. These two matrices are then added together to give the new stiffness matrix for step 3, and the structure is subjected to the incremental vector of loads $\{\delta q_3{}^\circ\}$. The vectors of displacements $\{\delta u_3{}^\circ\}$ due to this loading vector are calculated and added to $\{u_2{}^\circ\}$ to give the resulting displacement vector $\{u_3{}^\circ\}$ due to the total load at this stage, namely, $\{\delta q_1{}^\circ\} + \{\delta q_2{}^\circ\} + \{\delta q_3{}^\circ\}$.

This process is continued until the desired load is reached or the structure has failed.

N.B. In Table 9.1 the following notation is used:

$[K_o^\circ(u_i^\circ)] =$ Structural stiffness matrix in global coordinates immediately after step 'i'

$[K_G^\circ(u_i^\circ)] =$ system geometrical stiffness matrix in global coordinates immediately after step 'i'

$\{u_n^\circ\} = \sum_1^n \{\delta u_1^\circ\} =$ a vector of nodal displacements in global coordinates at the end of the loading cycle

$\{q_n^\circ\} = \sum_1^n \{\delta q_i^\circ\} =$ a vector of nodal forces in global coordinates at the end of the loading cycle.

9.2 GEOMETRICAL NON-LINEARITY

The problems we will restrict ourselves to in this section will be those concerning non-linear problems within the elastic limit, and, in particular, elastic instability.

For elastic instability, there is no material non-linearity, so that $[K_o^\circ(0)]$ in step 1 of Table 9.1 becomes $[K^\circ]$, namely, the ordinary stiffness matrix. If there are no built-in stresses, then $[K_G^\circ(0)]$ in Table 9.1 is null, so that

$$\{u_1^\circ\} = [K^\circ]^{-1}\{q^\circ\}, \tag{9.3}$$

where

$\{q^\circ\}$ is unknown.

If, however, the relative magnitude of the forces is known,

$$\{q^\circ\} = \lambda\{q^{\circ*}\},$$

where

λ = a constant to be determined

$\{q^{\circ*}\}$ = a vector containing the relative magnitudes of the applied forces required to cause buckling.

Hence,

$$\{u_1^\circ\} = [K^\circ]^{-1}\lambda\{q^{\circ*}\} \tag{9.4}$$

Putting $\lambda = 1$ in (9.4):

$$\{u_1^{\circ*}\} = [K^\circ]^{-1}\{q^{\circ*}\}, \tag{9.5}$$

so that $[K_G^\circ(u_1^\circ)^*]$ can be determined.

However,

$$[K_G^\circ(u_1^\circ)] = \lambda[K_G^\circ(u_1^\circ)^*]. \tag{9.6}$$

As the problem is one of elastic instability, $[K°]$ will not change, hence by substituting (9.6) into a modified form of (2.11),

$$\{q°\} = ([K°] + \lambda[K_G°(u_1°)^*]) \{u_i°\} \tag{9.7}$$

Letting

$$[K_G°(u_1°)] = [K_G] = \text{the geometrical stiffness matrix due to } \{q°^*\}$$

and

$$\{q_{cr}°\} = \text{the load vector } \{q°\}, \text{ just prior to elastic instability, then (9.7)}$$
$$\text{becomes:}$$

$$\{q_{cr}°\} = ([K°] + \lambda[K_G°]) \{u_i°\}. \tag{9.8}$$

At buckling, $\{u_1°\} + [\infty\}$,

therefore

$$|[K°] + \lambda[K_G°]| = 0, \tag{9.9}$$

where

$$[K_G°] = \Sigma[k_G°]$$

$$[k_G°] = \text{the elemental geometrical stiffness matrix in global coordinates}$$

$$= [\Xi]^T [k_G] [\Xi]$$

$$[k_G] = \text{the elemental geometrical stiffness matrix in local coordinates}$$

$$[\Xi] \text{ is a matrix of directional cosines.}$$

Equation (9.9) is an eigenvalue problem, where the roots are 'λ', and

$$\{q_{cr}°\} = \lambda_{cr}\{q°^*\},$$

$$\lambda_{cr} = \text{the minimum, in magnitude, non-zero value of '}\lambda\text{'}$$

9.2.1   Geometrical stiffness matrix $[k_G]$

Prior to applying the method, it will be necessary to derive the geometrical stiffness matrix for the appropriate element, and to achieve this it will be necessary to determine the effects of large deflections on the axial strain of each element.

Consider the beam of Fig. 9.1, undergoing large deflections in the 'y' direction, due to a compressive axial force 'F'.

Consider an element of length 'δs' suffering a deflection 'δv' over a distance 'δx', so that

$$(\delta s)^2 = (\delta v)^2 + (\delta x)^2$$

or

$$\delta s = \left[1 + \left(\frac{\delta v}{\delta x}\right)^2\right]^{\frac{1}{2}} \delta x.$$

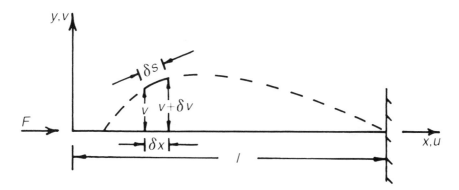

Fig. 9.1 – Large deflections of a beam.

In the limit,

$$\frac{ds}{dx} = = \left[1 + \tfrac{1}{2}\left(\frac{dv}{dx}\right)^2 - \tfrac{1}{8}\left(\frac{dv}{dx}\right)^4 + -\right].$$

If the term $(dv/dx)^4$ and other higher order terms are neglected, then the additional strain in the 'x' direction due to 'v'

$$= \delta\varepsilon_x = \tfrac{1}{2}\left(\frac{dv}{dx}\right)^2. \tag{9.10}$$

The total strain in the 'x' direction consists of the small deflection component of strain, namely, du/dx and the large deflection component, '$\delta\varepsilon_x$', i.e.

$$\varepsilon_x = \frac{du}{dx} + \tfrac{1}{2}\left(\frac{dv}{dx}\right)^2. \tag{9.11}$$

The effects of du/dx have already been taken into account in determining the stiffness matrix for the rod element of equation (3.4), and the combined effects of $\delta\varepsilon_x$ and du/dx will result in the derivation of the geometrical stiffness matrix $[k_G]$.

9.2.2 Rod element

The stiffness matrices for this element are of importance for pin-jointed trusses and are obtained by assuming the following displacement functions for 'u' and 'v':

$$u = u_1(1 - x/l) + u_2\, x/l$$

$$\tag{9.12}$$

$$v = v_1(1 - x/l) + v_2\, x/l$$

where u and v are as defined in Fig. 9.2.

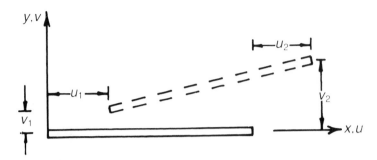

Fig. 9.2 – Rod element.

The strain energy stored in the bar

$$U_e = \frac{AE}{2} \int_0^l \varepsilon_x^2 \, dx.$$

Hence, from (9.11),

$$U_e = \frac{AE}{2} \int_0^l \left[\left(\frac{du}{dx}\right)^2 + \frac{du}{dx}\left(\frac{dv}{dx}\right)^2 + \frac{1}{4}\left(\frac{dv}{dx}\right)^4 \right] dx.$$

Neglecting $(dv/dx)^4$,

$$U_e = \frac{AE}{2} \int_0^l \left[\left(\frac{du}{dx}\right)^2 + \frac{du}{dx}\left(\frac{dv}{dx}\right)^2 \right] dx. \qquad (9.13)$$

Substituting the first derivatives of (9.12) into (9.13),

$$U_e = \frac{AE}{2} \int_0^l \left[\left(\frac{-u_1 + u_2}{l}\right)^2 + \left(\frac{-u_1 + u_2}{l}\right)\left(\frac{-v_1 + v_2}{l}\right)^2 \right] dx \qquad (9.14)$$

$$= \frac{AE}{2l}(u_1{}^2 - 2u_1 u_2 + u_2{}^2) + \frac{AE}{2l^2}(u_2 - u_1)(v_1{}^2 - 2v_1 v_2 + v_2{}^2)$$

$$= \frac{AE}{2l}(u_1{}^2 - 2u_1 u_2 + u_2{}^2) + \frac{F}{2l}(v_1{}^2 - 2v_1 v_2 + v_2{}^2), \qquad (9.15)$$

where

$$F = \frac{AE}{l}(u_2 - u_1),$$

the axial force in the rod. Applying the principle of minimum potential to each nodal displacement,

$$\frac{\partial U_e}{\partial u_1} = \frac{AE}{l}(u_1 - u_2)$$

$$\frac{\partial U_e}{\partial v_1} = \frac{F}{l}(v_1 - v_2)$$

$$\frac{\partial U_e}{\partial u_2} = \frac{AE}{l}(-u_1 + u_2)$$

$$\frac{\partial U_e}{\partial v_2} = \frac{F}{l}(-v_1 + v_2)$$

(9.16)

Rewriting (9.16) in the familiar stiffness displacement form,

$$[k]\{u_i\} = \left(\frac{AE}{l} \begin{bmatrix} 1 & 0 & -1 & 0 \\ 0 & 0 & 0 & 0 \\ -1 & 0 & 1 & 0 \\ 0 & 0 & 0 & 0 \end{bmatrix} + \frac{F}{l} \begin{bmatrix} 0 & 0 & 0 & 0 \\ 0 & 1 & 0 & -1 \\ 0 & 0 & 0 & 0 \\ 0 & -1 & 0 & 1 \end{bmatrix} \right) \begin{Bmatrix} u_1 \\ v_1 \\ u_2 \\ v_2 \end{Bmatrix}.$$

(9.17)

Comparing (9.17) with (9.1),

$$[k_o] = \frac{AE}{l} \begin{array}{cccc} u_1 & v_1 & u_2 & v_2 \\ \begin{bmatrix} 1 & 0 & -1 & 0 \\ 0 & 0 & 0 & 0 \\ -1 & 0 & 1 & 0 \\ 0 & 0 & 0 & 0 \end{bmatrix} & \begin{array}{c} u_1 \\ v_1 \\ u_2 \\ v_2 \end{array} \end{array}$$

and

$$[k_G] = \frac{F}{l} \begin{array}{cccc} u_1 & v_1 & u_2 & v_2 \\ \begin{bmatrix} 0 & 0 & 0 & 0 \\ 0 & 1 & 0 & -1 \\ 0 & 0 & 0 & 0 \\ 0 & -1 & 0 & 1 \end{bmatrix} & \begin{array}{c} u_1 \\ v_1 \\ u_2 \\ v_2 \end{array} \end{array}$$

where $[k_o]$ will be identical to the stiffness matrix of the rod of Chapter 3, providing 'E' remains constant, and $[k_G]$ is the corresponding geometrical stiffness matrix in local coordinates.

To obtain the geometrical stiffness matrix in global coordinates, consider the rod of Fig. 9.3. The geometrical stiffness matrix in global coordinates is given by

$$[k_G^o] = [\Xi]^T [k_G] [\Xi]$$

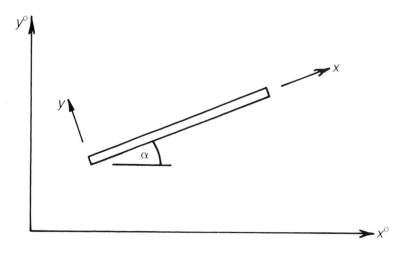

Fig. 9.3.

where

$$[\Xi] = \left[\begin{array}{c|c} \zeta & O_2 \\ \hline O_2 & \zeta \end{array}\right]$$

$$[\zeta] = \begin{bmatrix} c & s \\ -s & c \end{bmatrix}$$

$c = \cos \alpha, s = \sin \alpha$.

Hence, for a rod in global coordinates,

$$[k_G^o] = \frac{F}{l} \begin{bmatrix} s^2 & -cs & -s^2 & cs \\ -cs & c^2 & cs & -c^2 \\ -s^2 & cs & s^2 & -cs \\ cs & -c^2 & -cs & c^2 \end{bmatrix}. \tag{9.18}$$

It should be noted that the geometrical stiffness matrix is dependent on the internal force in the member, 'F'.

9.2.3 Beam element

The stiffness matrices for this element are useful for a number of problems, including that of the buckling of struts and frames and also for laterally loaded

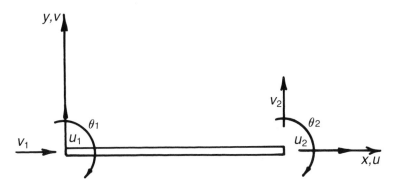

Fig. 9.4.

struts and ties, etc. In this case it will be necessary to combine the displacement function of a rod, with that of a beam in flexure, as follows:

$$u = u_1(1 - \xi) + u_2 \xi$$
$$v = v_1(1 - 3\xi^2 + 2\xi^3) + \theta_1 l(1 + 2\xi^2 - \xi^3) + v_2(3\xi^2 - 2\xi^3)$$
$$\qquad + \theta_2 l(\xi^2 - \xi^3), \tag{9.19}$$

where $\xi = x/l$.

It can readily be shown that the strain in the 'x' direction of any fibre at a distance 'y' from the neutral axis of the beam, is given by

$$\varepsilon_x = \frac{du}{dx} + \tfrac{1}{2}\left(\frac{dv}{dx}\right)^2 - \frac{d^2v}{dx^2}\,y, \tag{9.20}$$

where

$$\tfrac{1}{2}\left(\frac{dv}{dx}\right)^2 = \text{additional strain due to large deflections}$$

and

$$-\frac{d^2v}{dx^2}y = \text{direct strain, due to flexure, in a fibre at a distance '}y\text{' from the neutral axis in bending.}$$

Now,

$$U_e = \frac{E}{2}\int_{\text{vol}} \varepsilon_x{}^2 \, d(\text{vol}) = \text{strain energy in the beam.}$$

Hence, by substitution of the respective derivatives of (9.12) and (9.19) into (9.20), U_e can be determined. Neglecting the term $(dv/dx)^4$, and applying the principle of minimum potential, the following are obtained;

$$([k_o] + [k_G])\{u_i\} = \left\{ \frac{EI}{l^3} \begin{bmatrix} \dfrac{Al^2}{I} & & & & & \\ 0 & 12 & \text{Symmetrical} & & & \\ 0 & -6l & 4l^2 & & & \\ -\dfrac{Al^2}{I} & 0 & 0 & \dfrac{Al^2}{I} & & \\ 0 & -12 & 6l & 0 & 12 & \\ 0 & -6l & 2l^2 & 0 & 6l & 4l^2 \end{bmatrix} \right.$$

$$\left. + \frac{F}{l} \begin{bmatrix} 0 & & & & & \\ 0 & \dfrac{6}{5} & \text{Symmetrical} & & & \\ 0 & -\dfrac{l}{10} & \dfrac{2l^2}{15} & & & \\ 0 & 0 & 0 & 0 & & \\ 0 & -\dfrac{6}{5} & \dfrac{l}{10} & 0 & \dfrac{6}{5} & \\ 0 & -\dfrac{l}{10} & -\dfrac{l^2}{30} & 0 & \dfrac{l}{10} & \dfrac{2l^2}{15} \end{bmatrix} \right\} \begin{Bmatrix} u_1 \\ v_1 \\ \theta_1 \\ u_2 \\ v_2 \\ \theta_2 \end{Bmatrix} \quad (9.21)$$

where

$$F = \frac{AE}{l} \times (u_2 - u_1) = \text{constant} = \text{axial force in the beam}$$

$[k_G]$ = the geometrical stiffness matrix for a beam in local coordinates.

In terms of global coordinates, the geometrical stiffness matrix for a beam-column element is given by

$$[k_G^o] = \frac{F}{l} \begin{bmatrix} & u_1{}^o & v_1{}^o & \theta_1 & u_2{}^o & v_2{}^o & \theta_2 \\ \dfrac{6}{5}s^2 & & & & & & u_1{}^o \\ -\dfrac{6}{5}cs & \dfrac{6}{5}c^2 & & & & & v_1{}^o \\ \dfrac{l}{10}s & -\dfrac{l}{10}c & \dfrac{2l^2}{15} & & & & \theta_1 \\ -\dfrac{6}{5}s^2 & \dfrac{6}{5}cs & -\dfrac{l}{10}s & \dfrac{6}{5}s^2 & & & u_2{}^o \\ \dfrac{6}{5}cs & \dfrac{6}{5}c^2 & \dfrac{l}{10}c & -\dfrac{6}{5}cs & \dfrac{6}{5}c^2 & & v_2{}^o \\ \dfrac{l}{10}s & -\dfrac{l}{10}c & -\dfrac{l^2}{30} & -\dfrac{l}{10}s & \dfrac{l}{10}c & \dfrac{2l^2}{15} & \theta_2 \end{bmatrix} \quad (9.22)$$

9.2.4   Alternative method of finding $[k_G]$

This method is due to Martin [59], and is based on the method of *minimum potential*. Now,

Total potential energy = Strain energy
+ Potential energy of load system,

or

$$\pi_{\mathrm{p}} = \tfrac{1}{2}\{u_1\}^{\mathrm{T}}[k]\{u_i\} - \{u_1\}^{\mathrm{T}}\{p\},$$

where

$$U_{\mathrm{e}} = \tfrac{1}{2}\{u_i\}^{\mathrm{T}}[k]\{u_i\} = \text{strain energy.} \tag{9.23}$$

From (9.23) it can be seen that

$$k_{ij} = \frac{\partial^2 U_{\mathrm{e}}}{\partial u_i\,\partial u_j}. \tag{9.24}$$

Now for large displacements,

$$\varepsilon = \varepsilon_0 + \varepsilon_{\mathrm{a}}$$

where

ε_0 = initial strain at start of step

ε_{a} = additional strain at step.

N.B. For elastic instability, there is only one step.

Now,

$$U_{\mathrm{e}} = \frac{1}{2} \int\limits_{\mathrm{vol}} \sigma\varepsilon\,\mathrm{d(vol)},$$

and for a rod element,

$$U_{\mathrm{e}} = \tfrac{1}{2}AE \int\limits_0^l \varepsilon^2\,\mathrm{d}x$$

$$= \tfrac{1}{2}AEl\,\varepsilon_0{}^2 + AE\,\varepsilon_0 \int\limits_0^l \varepsilon_{\mathrm{a}}\,\mathrm{d}x + \tfrac{1}{2}AE \int\limits_0^l \varepsilon_{\mathrm{a}}{}^2\,\mathrm{d}x$$

$$= U_0 + U_1 + U_2. \tag{9.25}$$

From (9.11),

$$\varepsilon_{\mathrm{a}} = \frac{\mathrm{d}u}{\mathrm{d}x} + \tfrac{1}{2}\left(\frac{\mathrm{d}v}{\mathrm{d}x}\right)^2, \tag{9.26}$$

and substituting (9.12) into (9.25) and (9.26),

$$U_e = U_o + AE\,\varepsilon_o \int_0^l \left[\frac{(u_2 - u_1)}{l} + \frac{1}{2}\left(\frac{v_2 - v_1}{l}\right)^2\right]dx$$

$$+ \frac{1}{2}AE \int_0^l \left[\frac{(u_2 - u_1)}{l} + \frac{1}{2}\left(\frac{v_2 - v_1}{l}\right)^2\right]^2 dx. \tag{9.27}$$

Letting $F = AE\,\varepsilon_o$ = the initial loading, and noting from (9.24) that only quadratic terms need be retained, (9.27) simplifies to

$$U_e = \frac{1}{2}\frac{F}{l}(v_2 - v_1)^2 + \frac{1}{2}\frac{AE}{l}(u_2 - u_1)^2$$

$$= \frac{1}{2}\frac{F}{l}[v_1 \quad v_2]\begin{bmatrix} 1 & -1 \\ -1 & 1 \end{bmatrix}\begin{Bmatrix} v_1 \\ v_2 \end{Bmatrix}$$

$$+ \frac{1}{2}\frac{AE}{l}[u_1 \quad u_2]\begin{bmatrix} 1 & -1 \\ -1 & 1 \end{bmatrix}\begin{Bmatrix} u_1 \\ u_2 \end{Bmatrix}. \tag{9.28}$$

Comparing (9.28) with (9.23),

$$\begin{matrix} & v_1 & v_2 & \\ [k_G] = \frac{F}{l} & \begin{bmatrix} 1 & -1 \\ -1 & 1 \end{bmatrix} & \begin{matrix} v_1 \\ v_2 \end{matrix} \end{matrix} = \text{geometrical stiffness matrix}$$

and

$$\begin{matrix} & u_1 & u_2 & \\ \{k_o\} = \frac{AE}{l} & \begin{bmatrix} 1 & -1 \\ -1 & 1 \end{bmatrix} & \begin{matrix} u_1 \\ u_2 \end{matrix} \end{matrix} = \text{elemental stiffness matrix},$$

which can be seen to be of the same form as before for both cases.

9.2.5 Example 9.1

A plane truss is constructed from two rods of a constant AE of 8 MN, and are firmly pinned at 1 and 2 and to each other at 3, as shown in Fig. 9.5. Calculate the load W that will cause instability of the structure.

Element 2–3

   Length $= l; c = 1; s = 0$.

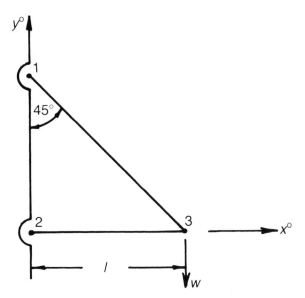

Fig. 9.5.

Hence, from (3.16),

$$[k^o]_{2-3} = \frac{AE}{l} \begin{array}{cccc} u_2^{\,o} & v_2^{\,o} & u_3^{\,o} & v_3^{\,o} \\ \begin{bmatrix} 1 & 0 & -1 & 0 \\ 0 & 0 & 0 & 0 \\ -1 & 0 & 1 & 0 \\ 0 & 0 & 0 & 0 \end{bmatrix} & \begin{array}{l} u_2^{\,o} \\ v_2^{\,o} \\ u_3^{\,o} \\ v_3^{\,o} \end{array} \end{array} \qquad (9.29)$$

From elementary statics or the matrix displacement method, it can be shown that

$$F_{2-3} = -W.$$

Hence, from (9.18),

$$[k_G^{\,o}]_{2-3} = -\frac{W}{l} \begin{array}{cccc} u_2^{\,o} & v_2^{\,o} & u_3^{\,o} & v_3^{\,o} \\ \begin{bmatrix} 0 & 0 & 0 & 0 \\ 0 & 1 & 0 & -1 \\ 0 & 0 & 0 & 0 \\ 0 & -1 & 0 & 1 \end{bmatrix} & \begin{array}{l} u_2^{\,o} \\ v_2^{\,o} \\ u_3^{\,o} \\ v_3^{\,o} \end{array} \end{array} \qquad (9.30)$$

Element 1–3

Length $= \sqrt{2}l$; $c = 0.707$; $s = -0.707$.

From (3.16),

$$[k^\circ]_{1-3} = \frac{AE}{2\sqrt{2}l} \begin{array}{cccc} u_1^\circ & v_1^\circ & u_3^\circ & v_3^\circ \end{array}$$

$$[k^\circ]_{1-3} = \frac{AE}{2\sqrt{2}l} \begin{bmatrix} 1 & -1 & -1 & 1 \\ -1 & 1 & 1 & -1 \\ -1 & 1 & 1 & -1 \\ 1 & -1 & -1 & 1 \end{bmatrix} \begin{array}{c} u_1^\circ \\ v_1^\circ \\ u_3^\circ \\ v_3^\circ \end{array} \qquad (9.31)$$

From elementary statics or the matrix displacement method, it can be shown that

$$F_{1-3} = \sqrt{2}\,W.$$

Hence, from (9.18),

$$[k_G^\circ]_{1-3} = \frac{W}{2l} \begin{bmatrix} 1 & 1 & -1 & -1 \\ 1 & 1 & -1 & -1 \\ -1 & -1 & 1 & 1 \\ -1 & -1 & 1 & 1 \end{bmatrix} \begin{array}{c} u_1^\circ \\ v_1^\circ. \\ u_3^\circ \\ v_3^\circ \end{array} \qquad (9.32)$$

Summing the terms of (9.29) and (9.31), corresponding to the free displacements, u_3° and v_3°, the following is obtained for the constrained system stiffness matrix:

$$[K^\circ_{11}] = \frac{AE}{l} \begin{bmatrix} 1 + \dfrac{1}{2\sqrt{2}} & -\dfrac{1}{2\sqrt{2}} \\ -\dfrac{1}{2\sqrt{2}} & \dfrac{1}{2\sqrt{2}} \end{bmatrix} \begin{array}{c} u_3^\circ \\ v_3^\circ. \end{array} \qquad (9.33)$$

Similarly, from (9.30) and (9.32), the system geometrical stiffness matrix corresponding to the free displacements, u_3° and v_3°, is as follows:

$$[K^\circ_{G11}] = \frac{W}{l} \begin{bmatrix} \frac{1}{2} & \frac{1}{2} \\ \frac{1}{2} & -\frac{1}{2} \end{bmatrix} \begin{array}{c} u_3^\circ \\ v_3^\circ. \end{array} \qquad (9.34)$$

Substituting (9.33) and (9.34) into (9.9), the eigenvalue problem becomes

$$\left| \frac{AE}{l} \begin{bmatrix} 1 + \dfrac{1}{2\sqrt{2}} & -\dfrac{1}{2\sqrt{2}} \\ -\dfrac{1}{2\sqrt{2}} & \dfrac{1}{2\sqrt{2}} \end{bmatrix} + \frac{W}{l} \begin{bmatrix} \frac{1}{2} & \frac{1}{2} \\ \frac{1}{2} & -\frac{1}{2} \end{bmatrix} \right| = 0,$$

or

$$(10.832 + 0.5W)(2.832 - 0.5W) - (-2.832 + 0.5W)^2 = 0$$
$$22.6 - 1.168W - 0.5W^2 = 0,$$

the smallest eigenvalue of which is

$$W = 5.66 \text{ MN} = \text{elastic instability load.}$$

It is interesting to note that the collapse load for this truss is about 30 per cent larger than that of the similar truss of Przemieniecki[45] (p. 397), which was subjected to a load parallel to 2–3 at 3. This is because in the truss of Przemieniecki, the member 1–3 had no load in it, and thus made a smaller contribution to the buckling strength of the structure.

9.2.6 Example 9.2

Calculate the axial buckling load for a strut which is fixed at one end and free at the other. Neglect the elements in the stiffness matrices corresponding to the deflection 'u'.

Fig. 9.6.

Assuming a one element idealisation and substituting into the relevant matrices of (4.13) and (9.22), but eliminating unwanted rows and columns,

$$[K^\circ_{11}] = \frac{EI}{l^3} \begin{array}{cc} v_2^\circ & \theta_2^\circ \\ \begin{bmatrix} 12 & 6l \\ 6l & 4l^2 \end{bmatrix} & \begin{array}{c} v_2^\circ \\ \theta_2^\circ \end{array} \end{array}$$

$$[K^\circ{}_{G11}] = \frac{F}{l} \begin{array}{c} v_2{}^\circ \quad \theta_2{}^\circ \\ \begin{bmatrix} \dfrac{6}{5} & \dfrac{l}{10} \\[2mm] \dfrac{l}{10} & \dfrac{2l^2}{15} \end{bmatrix} \end{array} \begin{array}{c} v_2{}^\circ . \\[6mm] \theta_2{}^\circ \end{array}$$

The stability determinant becomes

$$\left| \frac{EI}{l^2} \begin{bmatrix} 12 & 6l \\ 6l & 4l^2 \end{bmatrix} + F \begin{bmatrix} \dfrac{6}{5} & \dfrac{l}{10} \\[2mm] \dfrac{l}{10} & \dfrac{2l^2}{15} \end{bmatrix} \right| = 0,$$

which, on expansion, gives the following quadratic:

$$\frac{3l^2 F^2}{20} + \frac{26EIF}{5} + \frac{12E^2 I^2}{l^2} = 0,$$

therefore

$$F = -2.465EI/l^2,$$

or

Elastic buckling load $= 2.465EI/l^2$,

which is exactly the same as the well-known analytical value.

9.2.7 Example 9.3

Determine the maximum bending moment for the laterally loaded tie-bar of Fig. 9.7, using two equal length elements.

$$E = 2 \times 10^{11} \text{ N/m}^2 \qquad I = \text{2nd moment of area} = 1 \times 10^{-7} \text{ m}^4.$$

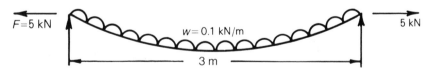

$F=5$ kN $w = 0.1$ kN/m   5 kN   3 m

Fig. 9.7 – Laterally loaded tie-bar.

From the computer program,

Maximum deflection $\hat{\delta}$ $= -4.298\text{E-}3$ m

Maximum bending moment $\hat{M} = -0.0877$ kN m

From the classical theory of laterally loaded tie-bars,

$$\hat{M} = -\frac{W}{\alpha^2}(1 - \operatorname{sech}(\alpha l/2)) = 0.091 \text{ kN m}.$$

9.2.8 Example 9.4

Determine the maximum bending moment for the laterally loaded strut of Fig. 9.18, using two equal length elements.

$$E = 2 \times 10^{11} \text{ N/m}^2, \qquad I = 1 \times 10^{-7} \text{ m}^4$$

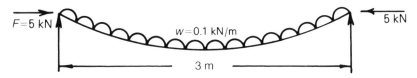

$F=5$ kN $w=0.1$ kN/m   5 kN

3 m

Fig. 9.8 – Laterally loaded strut.

From the computer program,

Maximum deflection $\qquad = \hat{\delta} = \underline{-6.8198\text{E-3 m}}$

Maximum bending moment $\quad = \hat{M} = \underline{-0.152 \text{ kN m}}$.

From the classical theory of laterally loaded struts,

$$\hat{\delta} = \frac{W}{F\alpha^2}\left(1 - \sec\frac{\alpha l}{2}\right) + \frac{Wl^2}{8F} = \underline{-6.836\text{E-3 m}}$$

$$\hat{M} = \frac{W}{\alpha^2}\left(1 - \sec\frac{\alpha l}{2}\right) = -0.147 \text{ kN m}.$$

The results for Examples 9.3 and 9.4 show good agreement between the finite element solution and classical theory, despite the fact that only two elements were used for the former. The results have also shown that a tensile axial load tends to increase bending stiffness, whereas a compressive axial load tends to decrease bending stiffness.

The main advantage of the computer solution over classical theory is that it can cater for struts and ties with complex combinations of lateral loads, and with complex boundary conditions, as shown in Fig. 9.9.

F F

Fig. 9.9 – Laterally loaded continuous tie-bar.

N.B. It should be noted that the effects of initial imperfections have not been included in this section, and that these can considerably decrease the theoretical buckling loads. Some such structures should theoretically fail through elastic instability, but because of initial imperfections, suffer elastic knockdown, and, as a result of this, fail at buckling loads which are only a small fraction of the theoretical predictions.

9.3 PLASTICITY

There are some problems in which the effects of material non-linearity are important, but those due to geometrical non-linearity are not. A typical example in this category is that of the plane stress problem which is involved with stress concentrations due to holes, etc. For cases such as these, it is usual to analyse the problem as shown in Table 9.1, but to neglect the effects of $[k_G]$. The method is a step-by-step procedure, where the effects of stress on tangent modulus are determined at the end of each step, and substituted into the next step, thus making the analysis linear over each step. One simple process of attaining this is to obtain the stress–tangent modulus relationship for the material, as shown in Fig. 9.10, and after calculating the stress in the elements throughout the structure, to determine their respective tangent moduli from this figure for the next step.

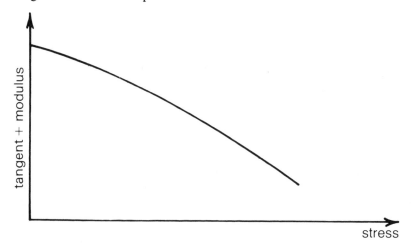

Fig. 9.10 – Tangent modulus-stress relationship.

The stresses and deflections of all the steps, up to the one considered, should be superimposed, and because of this it is evident that the process will, in general, overestimate the stiffness of the structure, as shown by the dotted line of Fig. 9.11. It should be noted that, for greater precision, it is a fairly simple matter to extend this approach to the shear stress criterion for yield.

The approach shown in this section is a very elementary one and does not include the effects of hardening and cyclic loading. These latter effects are beyond the scope of the present book, and the reader who is interested in such phenomena is referred to references [13 and 31].

9.4 NON-LINEAR STRUCTURAL VIBRATIONS

As any axial load in a member affects its bending stiffness, it will also affect its natural frequencies of vibration. In general, a compressive axial force on a

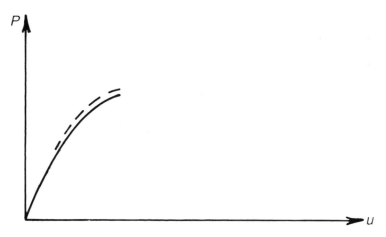

Fig. 9.11 – Load–displacement relationship.

structure will tend to decrease its natural frequencies of vibration, and a tensile axial force will tend to increase its natural frequencies of vibration.

In the present text, only non-linear elastic vibrations will be considered, so that the effect of initial stresses will be to modify the stiffness matrix through the geometrical stiffness matrix. The process, therefore, for allowing for the effects of tensile and compressive forces, on natural frequencies, is a long and involved one. It can be described as follows.

Stage 1
From

$$\{q^\circ\} = [K^\circ]\{u_1{}^\circ\}$$

determine $\{u_1{}^\circ\}$ and initial stresses.

Stage 2
Determine $[K_G{}^\circ]$, knowing the initial stresses from stage 1.

Stage 3
Set up the dynamical equations

$$|[K^\circ] + [K_G{}^\circ] - \omega^2[M^\circ]| = 0. \qquad (9.35)$$

Stage 4
Assuming

$$[K_2] = [K^\circ] + [K_G{}^\circ],$$

modify the dynamical equation of (9.35) to (9.36):

$$|[K_2] - \omega^2[M^\circ]| = 0, \qquad (9.36)$$

where

$\{q^\circ\}$ = a vector of applied nodal forces

$[M^\circ]$ = system mass matrix in global coordinates

$[K^\circ]$ = system stiffness matrix in global coordinates

$[K_G^\circ]$ = system geometrical stiffness matrix.

Equation (9.36) can now be solved by standard eigenvalue techniques. To demonstrate the method, the following examples will be considered.

9.4.1 Example 9.5

Determine the natural frequencies of vibration for a long slender strut of length 1 m, fixed at one end and free at the other, under the following axial loads:

(a) A compressive axial load of 20 kN
(b) A compressive axial load of 40 kN
(c) A tensile axial load of 40 kN
(d) Hence, or otherwise, determine the fundamental flexural frequency for this strut for various axial loads, varying from a tensile one of 50 kN to the theoretical elastic instability load.

The following properties relate to the strut:

$$E = 2 \times 10^{11} \text{ N/m}^2 \qquad \rho = 7860 \text{ kg/m}^3$$
$$I = \text{2nd moment of area} = 1 \times 10^{-7} \text{ m}^4$$
$$A = \text{cross-sectional area} = 1 \times 10^{-4} \text{ m}^2.$$

From (4.13),

$$[K_{11}] = \begin{array}{cc} v_2 & \theta_2 \\ \begin{bmatrix} 240\,000 & 120\,000 \\ 120\,000 & 80\,000 \end{bmatrix} & \begin{array}{c} v_2 \\ \theta_2. \end{array} \end{array}$$

From (7.41),

$$[M_{11}] = \begin{array}{cc} v_2 & \theta_2 \\ \begin{bmatrix} 0.292 & 0.0412 \\ 0.0412 & 7.486E\text{-}3 \end{bmatrix} & \begin{array}{c} v_2 \\ \theta_2. \end{array} \end{array}$$

(a) Substituting $F = 20\,000$ into (9.22),

$$[K_{G11}] = \begin{bmatrix} -24\,000 & -2000 \\ -2000 & -2667 \end{bmatrix}.$$

Hence, the dynamic equation becomes

$$\left| \begin{bmatrix} 240\,000 - 24\,000 & 120\,000 - 2000 \\ 120\,000 - 2000 & 80\,000 - 2667 \end{bmatrix} - \omega^2 \begin{bmatrix} 0.292 & 0.0412 \\ 0.0412 & 7.486\text{E-3} \end{bmatrix} \right| = 0,$$

which on expansion becomes,

$$4.89 \times 10^{-4}\,\omega^4 - 14475\,\omega^2 + 2.776 \times 10^9 = 0,$$

yielding

$$\omega_1 = 439.6 \text{ rads/s}$$
$$\omega_2 = 5423 \text{ rads/s}$$

and

$$n_1 = 69.97 \text{ Hz}; \quad \lfloor \theta_1 \quad \theta_2 \rfloor = \lfloor 1 \quad -1 \rfloor$$
$$n_2 = 863.1 \text{ Hz}; \quad \lfloor \theta_1 \quad \theta_2 \rfloor = \lfloor 1 \quad 1 \rfloor.$$

(b) $F = -40\,000$ N

$$\omega_1 = 253.9 \text{ rads/s}$$
$$\omega_2 = 5283 \text{ rads/s}$$

and

$$n_1 = 40.4 \text{ Hz}; \quad \lfloor \theta_1 \quad \theta_2 \rfloor = \lfloor 1 \quad -1 \rfloor$$
$$n_2 = 840.9 \text{ Hz}; \quad \lfloor \theta_1 \quad \theta_2 \rfloor = \lfloor 1 \quad 1 \rfloor$$

(c) $F = 40\,000$ N

$$\omega_1 = 744.8 \text{ rads/s}$$
$$\omega_2 = 5824 \text{ rads/s}$$
$$n_1 = 118.5 \text{ Hz}; \quad \lfloor \theta_1 \quad \theta_2 \rfloor = \lfloor 1 \quad -1 \rfloor$$
$$n_2 = 926.8 \text{ Hz}; \quad \lfloor \theta_1 \quad \theta_2 \rfloor = \lfloor 1 \quad 1 \rfloor,$$

where,

n_1 and n_2 = the first and second natural frequencies

$\lfloor \theta_1 \quad \theta_2 \rfloor$ = the eigenvector corresponding to n_1 or n_2.

(d) From the computer program, the results are given in Table 9.2, for the values of the fundamental flexural frequencies of vibration 'n_1', corresponding to the applied axial loads 'F', for this strut.

From the above results, it can be seen that the tensile axial loads cause the natural frequencies to increase in magnitude, and the compressive axial loads decrease the magnitudes of the frequencies.

Table 9.2. Variation of n_1 with F

F	n_1
50 000	124.5
40 000	118.5
30 000	112.2
20 000	105.3
10 000	97.87
0	89.69
− 10 000	80.52
− 20 000	69.97
− 30 000	57.27
− 40 000	40.40
− 45 000	28.22
− 47 500	19.38
− 49 000	11.04
− 49 500	6.09
− 49 700	1.81
− 49 710	1.25
− 49 720	0 ← Buckling load.

9.5 DYNAMIC INSTABILITY

Most studies of instability are based on static analysis, but it is possible that owing to periodic compressive forces, the fundamental frequency can so decrease that the structure can fail because of a load much smaller than the static instability prediction.

Problems in this category include structures such as that of a yacht mast and a submarine pressure hull. For the latter, as it descends deeper into the water, its resonant frequencies will decrease, until it reaches a point where the resonant frequencies are so low that any small exciting force can trigger off dynamic instability at a pressure considerably less than that which would have caused static instability.

In 1975 [61], the present author published a paper describing this phenomenon, and to illustrate this effect, he considered a thin-walled cylinder.

EXAMPLES FOR PRACTICE

For Examples (1) and (2), neglect the elements of the beam–column stiffness matrices, corresponding to the displacement 'u'.

1. Using a single element, calculate the elastic buckling load for a strut pinned at both ends.

   Answer: $-12EI/l^2$.

2. Using two equal length elements, calculate the elastic buckling loads for the following struts.

  (a) Fixed at both ends.

  Answer: $-40EI/l^2$.

  (b) Fixed at one end and pinned at the other.

  Answer: $-20.7EI/l^2$.

  (c) Fixed at both ends, as shown in Fig. 9.12

  $I_{1-2} = 10 \times 10^{-6}\,\text{m}^4\ I_{2-3} = 5 \times 10^{-6}\,\text{m}^4$.

  $E = 2 \times 10^8\,\text{kN/m}^2$

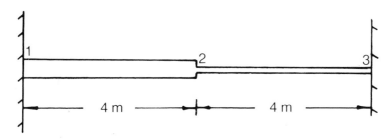

Fig. 9.12.

Answer: $-826.8\,\text{kN}$.

3. Calculate the theoretical buckling load 'p_{cr}' for the plane pin-jointed truss of Fig. 9.13, based on elastic instability assumptions.
  The members of the truss have the following properties:
  $A = 1 \times 10^{-4}\,\text{m}^2,\ E = 2 \times 10^{11}\,\text{N/m}^2$.

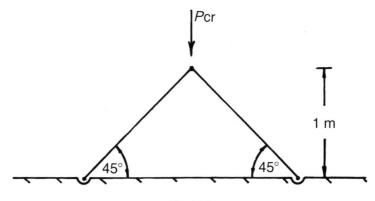

Fig. 9.13.

Answer: $-28.28\,\text{MN}$.

4. Calculate the elastic instability load, 'p_{cr}', for the truss of Example (3), assuming that this buckling load acts as shown in Fig. 9.14.

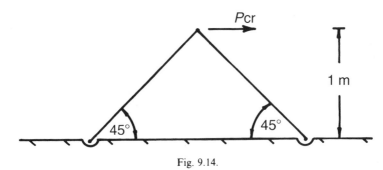

Fig. 9.14.

Answer: $P_{cr} = 28.28$ MN.

5. Calculate the theoretical buckling load 'P_{cr}' for the plane pin-jointed truss of Fig. 9.15, based on elastic instability assumptions.

$A = 1 \times 10^{-4}$ m^2, $E = 2 \times 10^{11}$ N/M^2

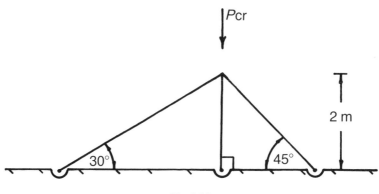

Fig. 9.15.

Answer: $P_{cr} = 0.581$ MN.

6. Calculate the lowest natural frequency of vibration for a beam of length 1 m pin-jointed at both ends and subjected to a compressive axial load of 100 kN, using only one element.

The following properties may be assumed to apply to the beam:

$A = 1 \times 10^{-4}$ m^2, $I = 1 \times 10^{-7}$ m^4, $E = 2 \times 10^{11}$ N/m^2,

$\rho = 7860$ kg/m^3

Answer: $n_1 = 278$ Hz.

Concluding remarks

The script has shown the power of the finite element method for analysing structures, particularly if they are of complex shape with complex boundary conditions. Providing one has access to a suitable computer, together with the necessary software associated with it, little difficulty will be experienced in analysing structures which defied theoretical analysis a few decades ago.

Much of the software written by the author, particularly that for use with microcomputers, can be quite easily used by engineers with little operational skills of these user-friendly machines. Users of these programs should, however, be thoroughly familiar with finite element methods, and this is one of the reasons why the present book has been written. Employers should avoid the temptation of using employees who are unfamiliar with engineering science and finite elements, as with most finite element packages much of the output is neatly packaged and 'very convincing'. In such cases, senior engineers can make quite ingenious interpretations of pages of 'rubbish', with disastrous consequences.

Users of such software should take care of avoiding numerical instability by joining a very stiff element to a very flexible one, or by choosing a badly shaped element. They should also take care of ensuring that the numerical precision of the machine has not been exceeded.

In fact, it is advisable that prior to applying any computer programs to practical problems, users should test these programs against known results, wherever this is possible. Particular care has to be taken of structural components in compression, as these can buckle owing to stresses which may be a small fraction of the stresses required to cause yield. There are many other factors which the engineer should also consider in his/her structural design, including stress concentrations, fatigue, and brittle fracture.

Through the use of variational calculus, the finite element method can also be applied to steady-state and transient field problems, including:

(a) The torsion of non-circular sections
(b) Fluid flow
(c) Heat transfer
(d) Electrostatics
(e) Magnetostatics
(f) Acoustics.

References

[1] Levy, S., Computation of Influence Coefficients for Aircraft Structures with Discontinuities and Sweepback, *J. Aero. Sci.*, **14**, 547–560, Oct., 1947.

[2] Levy, S., Structural Analysis and Influence Coefficients for Delta Wings, *J. Aero. Sci.*, **20**, 449–454, July, 1953.

[3] Argyris, J. H., Energy Theorems and Structural Analysis, *Aircraft Eng.*, Oct.-Nov., 1954 and Feb.-May, 1955.

[4] Willis, J., *Peanut Butter and Jelly Guide to Computers*, Dilthium Press, 1978.

[5] Turner, M. J., Clough, R. W., Martin, H. C. and Topp, L. J., Stiffness and Deflection Analysis of Complex Structures, *J. Aero. Sci.*, **23**, 805–823, 1956.

[6] Zienkiewicz, O. C. and Cheung, Y. K., Finite Element Analysis for Arch Dams and Comparison with Finite Difference Procedures, *Int. Symp. on Theory of Arch Dams*, Southampton, U.K., 123–140, 1964.

[7] Ergatoudis, J. G., Irons, B. M. and Zienkiewicz, O. C., Curved Isoparametric Quadrilateral Elements for Finite Element Analysis, *Int. J. Solids Struct.*, **4**, 31–42, 1968.

[8] Argyris, J. H., Triangular Elements with Linearly Varying Strain for the Matrix Displacement Method, *J. Roy. Aero. Soc., Tech. Note*, **69**, 711–713, Oct., 1965.

[9] Fraeijs de Veubeke, B., Displacement and equilibrium models in the finite element method, Chapter 9, *Stress Analysis*, ed. O. C. Zienkiewicz and G. Holister, Wiley, 1965.

[10] Clough, R. W., The Finite Element Method in Plane Stress, *Proc. 2nd A.S.C.E. Conf. on Electronic Computation*, Pittsburg, U.S.A., Sept., 1960.

[11] Zienkiewicz, O. C. and Cheung, Y. K., Buttress Dams on Complex Rock Foundations, *Water Power*, **16**, 193, 1964.

[12] Melosh, R. J., Basis for Derivation of Matrices for the Direct Stiffness Method, *J.A.I.A.A.*, **1**, 1631–1637, 1965.

[13] Zienkiewicz, O. C., *The Finite Element Method in Engineering Science*, McGraw-Hill, 1971.

[14] Courant, R., Variational Methods for the Solution of Problems of Equilibrium and Vibration, *Bull. Am. Math. Soc.*, **49**, 1–23, 1943.

[15] Ross, C. T. F., *Computational Methods in Structural and Continuum Mechanics*, Ellis Horwood, 1982.

[16] Ross, C. T. F., *Finite Element Programs for Axisymmetric Problems in Engineering*, Ellis Horwood, 1984.

[17] Ross, C. T. F., Microcomputer Applications to Structural and Continuum Mechanics, *Int. J. of Mech. Eng. Ed.*, **12**, 79–93, 1984.

[18] Irons, B. M., The Patch Test for Engineers, *Proc. Finite Element Symp.*, *Atlas Computing Lab.*, *Oxfordshire, U.K.*, 171–192, March 1975.

[19] Irons, B. and Ahmad, S., *Techniques of Finite Elements*, Ellis Horwood, 1980.

[20] Frazer, R. A., Duncan, W. J. and Collar, A. R., *Elementary Matrices and some Applications to Dynamics and Differential Equations*, Cambridge Univ. Press.

[21] Gere, J. M. and Weaver, W., *Matrix Algebra for Engineers*, Van Nostrand Reinhold, 1965.

[22] Jennings, A., *Matrix Algebra for Engineers and Scientists*, Wiley, 1977.

[23] Alexander, J. M., *Strength of Materials*-Vol. 1: Fundamentals, Horwood, 1981.

[24] Wilson, L. B., Solution of Certain Large Sets of Equations on Pegasus using Matrix Methods, *The Computer Journal*, **2**, No. 3, 130–133, 1959.

[25] Marshall, W. T., *Solution of Problems in Structures*, Pitman, 1964.

[26] Richards, T. H., *Energy Methods in Stress Analysis*, Ellis Horwood, 1977.

[27] Ross, C. T. F. and Johns, T., *Computer Analysis of Skeletal Structures*, SPON, 1981.

[28] Argyris, J. H., *Recent Advances in Matrix Methods of Structural Analysis*, Pergamon Press, 1964.

[29] Petyt, M., The Application of Finite Element Techniques to Plate and Shell Problems, *ISVR Report No. 120*, Feb., 1965.

[30] Timoshenko, S. P. and Goodier, J. N., *Theory of Elasticity*, 3rd Ed., McGraw-Hill Kogakusha Limited, 1970.

[31] Ford, H. and Alexander, J. M., *Advanced Mechanics of Materials*, 2nd Ed., Ellis Horwood, 1977.

[32] Kendrick, S., The Structural Design of Supertankers, *Trans. R.I.N.A.*, **112**, 391–420, 1970.

[33] Ross, C. T. F., Axisymmetric Finite Elements for Circular Plates and Cylinders, *Trans. R.I.N.A.*, **116**, 225–282, 1974.

[34] Scheid, F., *Numerical Analysis*, Schaum Outline Series.

[35] Timoshenko, S. P. and Young, D. H., *Theory of Structures*, McGraw-Hill Kogakusha Limited, 1965.

[36] Clough, R. W. and Tocher, J. L., Analysis of Thin Arch Dams by the Finite Method, *Int. Symp. on the Theory of Arch Dams, Southampton, U.K.*, 1964.

[37] Cantin, G. and Clough, R. W., A Curved Cylindrical Shell, Finite Element, *J.A.I.A.A.*, **6**, 1057, 1968.

[38] Ashwell, D. G. and Gallagher, R. H., (Editors), *Finite Elements for Thin Shells and Curved Members*, Wiley, 1976.

[39] Cook, R. D., *Concepts and Applications of Finite Element Analysis*, 2nd Ed., Wiley, 1974.

[40] Irons, B., Structural Eigenvalue Problems: Elimination of Unwanted Variables, *J.A.I.A.A.*, **3**, 961, 1965.

[41] Segerlind, L. J., *Applied Finite Element Analysis*, Wiley, 1976.

[42] Fenner, R. T., *Finite Element Methods for Engineers*, MacMillan, 1975.

[43] Irons, B. B., Engineering Applications of Numerical Integration in Stiffness Methods, *J.A.I.A.A.*, **4**, 2035–2037, 1966.

[44] Wylie, C. R., *Advanced Engineering Mathematics*, McGraw-Hill Kogakasha.

[45] Przemieniecki, J. S., *Theory of Matrix Structural Analysis*, McGraw-Hill, 1968.

[46] Ross, C. T. F., Vibration of a Model Tower, *Int. Conf. on Computer Applications in Civil Engineering, Roorkee, India*, III-51-III 54, Oct., 23–25, 1979.

[47] Bazeley, G. P., Cheung, Y. K., Irons, B. M. and Zienkiewicz, O. C., Triangular Elements in Bending Conforming and Non-conforming Solutions, *Proc. Conf. Matrix Methods in Structural Mechs., Wright-Patterson Air Force Base, Ohio*, AFFDL-TR-66-80, 547–576, 1966.

[48] Ross, C. T. F., Partially Conforming Plate Bending Elements for Static and Dynamic Analyses, *J. Strain Analysis*, **8**, 260–263, 1973.

[49] Guyan, R. J., Reduction of Stiffness and Mass Matrices, *J.A.I.A.A.*, **3**, 380, 1965.

[50] Barton, M. W., Vibration of Rectangular and Skew Cantilever Plates, J. Appl. Mech., *Trans. Am. Soc. Mech. Engrs.*, **73**, 129–134, 1951.

[51] Ross, C. T. F., Free Vibration of Thin Shells, *J. Sound and Vib.*, **39**, 337–344, 1975.

[52] Vedeler, G., The Distribution of Load in Longitudinal Strength Calculations, *Trans. R.I.N.A.*, **89**, 16–31, 1947.

[53] Holman, D. F., A Finite Series Solution for Grillages under Normal Loading, *Aero. Quarterly*, **7**, 49–57, 1957.

[54] Clarkson, J., *Data Sheets for the Elastic Design of Flat Grillages under Uniform Pressure, European Shipbuilding*, No. 8, 1959.

[55] Ross, C. T. F., Elastic Analysis of a Hatch Cover Grill, *The Shipbuilder and Marine-Engine Builder*, 158–160, April, 1964.

[56] Venancio-Filo, F. and Iguti, F., Vibrations of Grids by the Finite Element Method, *Computers and Structures*, **3**, 1331–1344, Pergamon Press, 1973.

[57] Turner, M. J., Dill, E. H., Martin, H. C. and Melosh, R. J., Large Deflections of Structures Subjected to Heating and External Loads, *J. Aero. Sci.*, **27**, 97–106 and 127, 1960.

[58] Turner, M. J., Martin, H. C. and Weikel, R. C., Further Development and Applications of the Stiffness Method, *AGARD ograph*, Pergamon Press, **72**, 203–266, 1964.

[59] Martin, H. C., On the Derivation of Stiffness Matrices for the Analysis of Large Deflection and Stability Problems, Proc. Conf. Matrix Methods in Structural Mechs., Wright-Patterson Air Force Base, Ohio, U.S.A., AFFDL-TR-66-80, 697–716, 1966.

[60] Argyris, J. H., Kelsey, S. and Kamel, H., Matrix Methods of Structural Analysis, *AGARD ograph*, Pergamon Press, **72**, 1–164, 1963.

[61] Ross, C. T. F., Finite Elements for the Vibration of Cones and Cylinders, *I.J.N.M.E.*, **9**, 833–845, 1975.

Index